Mystery in the Skies: UFOs in Fact, Fiction, and Folklore

ThisBookPress 2018

Also by David Rogers:
Thor's Hammer: A Novel
and
D. B. Cooper Is Dead:
A Solomon Starr Adventure

Available from Amazon in Print or Kindle

© David P. Rogers 2018

Front cover
M31--Andromeda Galaxy
Public domain

Back Cover
Sketch of Moses S. Coles' "Aerial Vessel" (US Patent No. 352298, November 9, 1886) from https://commons.wikimedia.org/wiki/Category:Airships_wave_of_1896-1897#/media/File:Cole_Air_Vessel_1897.jpg

Contents

1. The Problem with UFOs: *U* Is for *Unidentified*
2. The Really Strange Cases
3. Roswell, or How a Lot Was Made of Very Little
4. Interstellar Travel for Fun and Profit, or, How to Build Your Own UFO
5. UFOs And The Case For Scientific Uncertainty
6. Selected Other Cases More Convincing Than Roswell
7. Speculation: The Message

The Problem with UFOs: *U* is for *Unidentified*

The subject of UFOs has been for too long hijacked by opposing camps. On one side are the so-called skeptics. These are often better referred to as debunkers, who are misguided in their determination to accept any attempted mundane explanation of a UFO sighting, even if it does not actually explain the evidence. Planets, meteors, and ball lightning are the first resorts for pseudo-explanations; if these phenomena do not do the trick, there are always balloons and swamp gas. Failing all else, hoaxes and hallucinations figure prominently in the standard debunker's conclusions.

Those who approach the topic of UFO

sightings thus, with their minds already made up and a standard set of forgone conclusions to choose from, are not true skeptics. As Marcello Truzzi has pointed out, the term "skeptic" is much misused in UFO discussions. A real, honest skeptic does not hesitate to say "I don't know," or perhaps, "I don't know, and I doubt that anyone else does." True skepticism withholds judgment when the evidence is inadequate. Thus, those who rush to declare every UFO report a hoax or a simple case of misidentification of a star or planet are not skeptics. They are people with a belief system (implied but usually not stated) that everything can be explained by proper scientific investigation using known principles of physics. This approach is as unfounded in the facts as the attitude of the wild-eyed true believer, and equally far from real skepticism.

At the opposite extreme are true believers, for whom the imminent alien invasion is evident from every light in the sky. For the true believer, even sightings obviously explained as ordinary aircraft or meteors must be labeled as UFOs. Every mutilated cow and every wild tale told by publicity seekers will be taken as signs of extraterrestrial activity, even if the alleged witnesses admit they were intoxicated at the time and have no relevant training, experience, or expertise in observing the sky and distinguishing the ordinary from the mysterious. Many witnesses know nothing about astronomy and little about ordinary aircraft, yet well-intentioned UFO investigators, in their desire to find something unusual, have been misled by confused or fabricated accounts. For the misguided true believer, proof may come in the form of argument by innuendo, huge quantities of irrelevant evidence, and the testimony of unreliable or uninformed witnesses.

Both sides in these rhetorical battles often become emotional and accuse the other of distorting the truth. But there is a way to avoid these pitfalls. The third, middle way, the cautious path I try to steer, is to embrace the mystery. This is the way of true skepticism. The true skeptic does not forget that the *U* in *UFO* stands for *Unidentified*. Many reports can be rationally explained by known phenomena, but in some cases, mysteries remain. In the cases of true UFOs, the best thing we can say is, We do not know what was seen. To say an the object was unidentified and therefore must be of alien origin is to forget what the word "unidentified" means. In a few cases, there is strong evidence of something strange, not only from credible witnesses but from physical traces or effects left from the encounter. These are the most interesting, mysterious situations I will focus on here.

The famous, convincing cases (and one unfortunately famous but very unconvincing case) reviewed here are not the only ones worth studying or investigating, of course. For instance, the National UFO Reporting Center (NUFORC) has thousands of reports on file, available for anyone to see online. Most of these probably have mundane explanations. But if even one of them represents a real, unknown object, that case should be studied.

Thus, I am convinced of three points. First, UFOs are real (emphasis on the U--for Unidentified); second, UFOs are rare (I've never seen one), and third, UFOs and UFO lore are worth studying, not only for what we can learn about human nature, but because in a few cases a real, hard-to-identify object was almost certainly present. These are very different subjects of study, but equally important.

My purpose in this book is to show that UFOs exist and to examine what I consider the most convincing cases. I do not propose to say what UFOs are or or not, other than to note that *U* stands for *Unidentified*. If I seem to overemphasize that point, I do so because it is too often forgotten. So when I say UFOs are real or that they exist, I mean simply that we have good reason to believe a few things--real objects--have been seen and have not been satisfactorily explained. To reiterate, "Unidentified" is not a synonym for "alien" or "extraterrestrial" anymore than it is a synonym for "hoax."

The most interesting cases involve instances in which reliable witnesses reported sightings that cannot be convincingly explained as ordinary aircraft or astronomical or meteorological phenomena. In addition, the most puzzling cases leave physical evidence on the ground or leave lasting physical effects on witnesses, and the cases were investigated and documented soon afterward by knowledgeable, reliable investigators. In addition, the cases stand the test of time as unexplained phenomena. No good, definite, demonstrably true explanation has been put forth.

In these cases, it is possible to give satisfactory answers to these questions: When and where was the UFO reported, and by whom? How do we know about the witness(-es) who described what was seen? Who investigated, and what were the investigators' credentials, knowledge, or experience? What kind of evidence was left on the ground, and was the physical evidence available for inspection by various qualified people? Or was the evidence allegedly spirited away by government or military officials, like the rumored alien bodies and spaceship at Roswell, never to be seen again?

Without good answers to these questions, we are in the realm of folklore and myth--which is not to say that cases of myth and folklore are not interesting and worthy of study. Popular cases such as Roswell (where physical evidence and convincing first-hand witnesses to actual inexplicable events are remarkably scarce or non-existent, given how famous the case is) are worth studying, if only for what they reveal about human psychology. Jacques Vallee has done much valuable work in this area. Yet the study of myth and folklore and the study of physical events are separate enterprises.

One criterion that has often been cited for what makes a convincing case is photographic evidence. In the area of UFO research, however, a picture may not be worth a thousand words. In fact, a photograph is worth nothing at all unless there are also credible witnesses on record who verify that the photograph actually captures an image of the UFO they saw. Even in the pre-digital age, it was simply too easy to fake a photograph of an alleged UFO.

If you ask a dozen people who know something about UFOs for the most convincing cases of unidentified objects, you will probably get a dozen different answers, no doubt with quite a bit of overlap. I certainly will not claim that no other cases are as convincing as the ones I present here. However, many would include the Roswell incident, no doubt the most famous case ever, as it has become synonymous with UFOs, as a most convincing case. I do not find it so, as explained below. Fame and evidence are not the same things. Similarly, enough has already been said about Kenneth Arnold and the unidentified objects that flew like "saucers." (For any who have not heard/read that story (or want to see it one more time), it is only a click away.) For anyone who disagrees with my choice of most convincing cases, I say, to the computer! Do your research, and have your say. If the study of UFOs proves anything, it shows there's always room for another opinion and another book, podcast, or blog post.

Sources

Truzzi, Marcello. "On Pseudo Skepticism." https://sites.google.com/view/ufoskepticorg/on-pseudo-skepticism-by-marcello-truzzi. Accessed 22 July 2018.

The National UFO Reporting Center. http://www.nuforc.org/. Accessed 28 July 2018.

The Really Strange Cases

The following cases of UFO sightings seem to have particularly strong evidence, both in the form of eyewitnesses and physical evidence confirmed and documented by multiple individuals, with no obvious mundane explanation for what happened. To say they are unexplained does not mean that they are necessarily instances of alien visitation. But unanswered questions remain.

RENDLESHAM FOREST

The Rendlesham Forest incident is one of the most fascinating cases in recent UFO history, and is sometimes referred to as "Britain's Roswell." This is an inexact comparison at best, as the evidence that something truly inexplicable happened in the Rendlesham Forest case is much stronger and more convincing than in the case of Roswell. The Rendlesham incident involves multiple sightings by enlisted men and high ranking officers of the US Air Force, imprints on the ground that are often thought to be the marks left by the landing gear of a triangular craft, and above normal radiations reading from the alleged landing sight. The initial investigation and response by US Air Force personnel were rapid, thorough, and well-documented, and many of the witness have not been reluctant to speak about what they saw or to insist that it was out of the ordinary.

The story began in late December, 1980. Over the course of several nights, unusual lights and at least one solid object on the ground were reported by US Air Force personnel in Rendlesham Forest, not far outside the Woodbridge Royal Air Force Base in Suffolk County, England. The Woodbridge RAF Base and the nearby Bentwaters RAF Base were essentially controlled and operated by the U.S. under agreements with the British Ministry of Defense. Thus, the discussion of this incident centers largely on what was seen and done by the commissioned officers and enlisted men of the USAF.

Some confusion seems to have existed about the exact dates, mainly due to a minor error on the date in a notebook belonging to Sergeant Jim Penniston and a report filed by Lieutenant Colonel Charles Halt, the Deputy Base commander. Penniston and Halt, along with others, investigated and saw things they could not immediately identify in the woods that December. Too much has probably been made of the question of dates and chronology in this incident, as the question of what was seen is much more interesting that whether it was seen on the 26th or 27th of December. The focus on dates may have been fueled at least in part by debunkers who hope to uncover some contradiction and thus reduce the credibility of the witnesses. As many people often find themselves off by a day in their recollection of dates, however, more than this sort of mostly irrelevant detail must be taken into consideration to determine witnesses' credibility.

In any case, it now seems well established that the first incident took place early Friday morning, December 26, before dawn. Two enlisted men, Airman John Burroughs Sergeant Jim Penniston, went into the woods to investigate what was at first suspected to be a crashed airplane. Penniston's account includes seeing, at close range, an object of unknown type and origin which he believed to be a machine or artificial object, often described as a triangular craft. Penniston said he had been close enough to touch the object and observe what were believed to be symbols or hieroglyphs, which no one has been able to interpret. Suffolk police officers also visited the area in the woods and observed what they described as marks left by a landed craft, though they reported seeing no object. However, some officers speculated that the marks on the ground could have been made by animals.

Plaster casts were made of the marks in the ground at the alleged landing sight, and the castings that have since appeared at UFO conferences and photographs, which are claimed to have been made at the time, show somewhat conical formations. The castings would be more convincing if they were perfect triangles or rectangles, but they do, at least, provide some physical evidence that something happened in Rendlesham Forest that December, assuming they are authentic and not the record of holes made by rabbits or burrowing animals. Lieutenant Colonel Halt, who investigated in person on the second night, has said publicly that he has casts that were made of the marks, and he is a very believable witness. From a skeptical point of view, it makes much more sense to question what caused the marks than to question Halt's veracity.

Unusual activity was again reported on Friday night, leading to the investigation by Halt, on the night of the 26th and into the early morning of the 27th. He was accompanied into the woods by Sergeant Monroe Nevels, who brought a Geiger counter, and others, including at least one photographer. If any of the photographs showed anything truly extraordinary, such as the actual UFO, they were never made public. Halt, however, made a tape recording of his observations and conversation during the night, while he was in the woods, and discussion of increased radiation levels can be heard on the tape. Copies and transcripts of this recording have been made widely available in the Internet. The tape is one of the most fascinating and convincing pieces of evidence in this case.

In addition to discussion of the Geiger counter readings, the tape conveys Halt's sense of amazement as unidentified lights move back and forth and approach him and the other men in the woods that night. At one point Halt says, excitedly, "Here he comes from the south. He's coming toward us now!" It is difficult, if not impossible, to listen to the tape or read a transcript of it and not see that Col. Halt felt something truly remarkable was going on at the time. If he and the other Air Force personnel had been Boy Scouts on their first overnight camping trip, one could imagine they were confused and spooked by ordinary events. But these were military professionals whose training and experience would have precluded such naive reactions. They would not have been on their first trip to the woods after dark.

Besides the radiation readings that were reported to be slightly higher than usual, slash marks or abrasions on tree trunks and broken branches on the trees were seen in the alleged landing sight. However, many have interpreted the marks on the trees as indicators left by foresters to show that the marked trees should be cut, and as anyone who has ever visited the woods (or picked up branches from the lawn after a storm) knows, branches sometimes fall from trees for various reasons; in themselves, a few broken branches are unremarkable. On the tape made in the woods that night, Halt himself seems to express doubt as to how recent the marks on the tree trunks were, so there is reason to question how much, if anything, the abrasions had to do with the unusual events in the woods that December. Samples of the sap taken from the trees are also mentioned on the tape, but little of interest seems to have resulted from those samples, or if so, the results were not made public.

As for the radiation readings, though they were apparently slightly elevated at the time, the exact cause of the elevation is difficult to establish. Perhaps the most curious aspect of the radiation readings is that in at least one case, as heard on the Halt tape, the radiation was higher than normal on one side of the tree but not the other. If the trees' radiation levels were the result of some natural phenomenon such as contamination from groundwater, one would expect it to be more or less uniform around the whole trunk. However, many mundane sources of radiation exist--visitation by time travelers or extraterrestrial craft are not among the most likely sources.

The Halt tape indicates that the increased energy and radiation readings seemed to surround a central location inside three indentations in the ground, often assumed to have been made by the artificial object seen earlier. Others have claimed the marks could have been made by animals. Photographs of the area were taken, but at least some of the developed film reportedly was cloudy and showed nothing meaningful. The photographic failures, like much else in UFO studies, yields contradictory interpretations. If the film was in fact clouded, this might be the result of increased radiation in the area. On the other hand, some might suppose the photographs were of good quality but were suppressed because they revealed too much. Or possibly they simply fell victim to incompetent processing, as the proper handling of photographic film is an art unto itself. Thus, the nature of the alleged landing site remains, like much else in this case, a mystery.

To try to explain the lights seen in the forest, various possibilities, none of them very convincing, have been put forth. Some suggest the nearby lighthouse at Orford Ness was the source of the light, but the Air Force personnel were familiar with the lighthouse and its location and would not have been confused by it for long. In addition, the Halt tape records his observation that the light was moving, which of course rules out a stationary lighthouse. Other explanations include lights from trucks or machines from nearby farms, but it is hard to believe the witnesses would not have recognized ordinary automotive headlights, even if such vehicles had somehow managed to move silently through the woods without leaving automotive tire tracks.

Similarly, various Air Force personnel, especially when additional sightings occurred on following night, were reportedly attracted by the commotion, though no one had asked or ordered them to investigate. Their flashlights may have resulted in some unexpected lights in the forest. However, the experienced military personnel, such as Col. Halt and the sergeants, would surely have recognized flashlights. Again, these were not the first times they had been in the woods at night. Attempted explanations involving astronomical phenomena such as stars, planets, or meteors have been put forth, but of course they would have appeared in the sky, and thus do not explain the sightings on the ground in the woods.

Could the sightings have been the result of some military technology about which not even those high in the chain of command, such as Colonel Ted Conrad, the Base Commander who sent Halt into the woods to investigate, was informed? Possibly. People charged with keeping secrets may keep them too well at times, and thus draw more attention than if information were discreetly shared. Or, if someone transporting an experimental device had to make an unplanned stop for whatever reason, the woods near the two bases might have seemed a relatively safe place to do so.

Decades after the incident, controversy continues. As late as July of 2015, new reports regarding what happened continue to emerge. Lieutenant Colonel Halt insists that facts were concealed by higher officials, and in 2015 Halt said retired radar operators had told him that unusual objects were tracked on radar in December 1980, but that they had feared to make these observations public at the time. Conversely, Colonel Conrad, Halt's immediate superior in December 1980, denies strongly any official deception.

Much debunking effort has also focused on later statements made by Sergeant Penniston, which to many sound more like science fiction than solid evidence of a UFO encounter. Some time after the sighting, Penniston wrote in his notebook a series of ones and zeroes which he interpreted as binary code. He believed the numbers were somehow communicated telepathically to him as a result of the contact with the strange object in the forest. He also apparently came to believe that the craft was s sort of time machine, something sent by humans from the future, though for what purpose seems unclear.

The lack of any credible witnesses would certainly damage the case for any UFO sighting, but one witness whom many doubt does not detract from what other witnesses report. Whether one finds Penniston credible or not, the fact is that several other credible witnesses saw strange things. Some sort of secret apparatus of human construction, perhaps military, seems more plausible than a craft of extraterrestrial origin, but no records or evidence have been produced to show that any experimental military hardware was present. Thus, the exact nature and origin of what happened remains a mystery. Something very odd took place in Rendlesham Forest that December. I do not know what was seen, and I am unconvinced that anyone else does, either.

U stands for *Unidentified*.

Sources

Clarke, David. "New Light on Rendlesham." https://drdavidclarke.co.uk/secret-files/secret-files-4/. Accessed 4 Feb 2018.

"Physical Evidence Col Halt Has from the Rendlesham Forest Incident." https://www.youtube.com/watch?v=3y-8goWqacM. Accessed 5 Feb 2018.

Pope, Nick, with John Burroughs and Jim Penniston. (2014). *Encounter in Rendlesham Forest: The Inside Story of the World's Best-Documented UFO Incident*. New York: St. Martin's, 2014.

Ridpath, Ian. "Jim Penniston's Notebook." http://www.ianridpath.com/ufo/pennistonnotebook.htm. Accessed 4 Feb 2018.

"Rendlesham Forest."
http://www.nickpope.net/rendlesham-forest.htm. Accessed 5 Feb 2018.

"Rendlesham Forest UFO Sighting 'New Evidence' Claim."
http://www.bbc.com/news/uk-england-suffolk-33447592. Accessed 3 Feb 2018.

"Rendlesham Incident: US commander speaks for the first time about the 'Suffolk UFO.'"
http://www.telegraph.co.uk/news/newstopics/howaboutthat/ufo/8685868/Rendlesham-Incident-US-commander-speaks-for-the-first-time-about-the-Suffolk-UFO.html. Accessed 6 Feb 2018.

"The Halt Tape" https://www.youtube.com/watch?v=35ZVA7NE0iI. Accessed 2 Feb 2018.

THE CASE OF STEFAN MICHALAK, OR THE FALCON LAKE INCIDENT

The case of Stefan Michalak, which began in May of 1967, is fascinating, even if one believes those who claim it was hoax. If it was indeed a hoax, which seems most unlikely, it ranks high among the best-staged and most persuasive hoaxes. If events occurred as Michalak reported, the case ranks among the most remarkable close encounters with a craft of unknown origin. Either way, it is well worth consideration, for what it reveals of psychological or physical reality.

Michalak, on the weekend of May 19th and

20th, 1967, was exploring the wilderness near Falcon Lake in the Province of Manitoba, Canada. He is often described as an amateur geologist or prospector, and his interest in rocks and minerals seems clear. Some time near noon on the 20th, Michalak said, he saw what seemed to be two cigar- or cigarette-shaped objects in the sky, but as they approached he saw they appeared to be more round and saucer- or disc-shaped. One of the objects landed, while the other flew away. Michalak said he looked for insignia or letters, such as NASA, to show where the craft came from, but saw none.

Michalak later reported that he began to sketch the object when it landed, noting that it had at first looked orange or red but later appeared gray. He then approached the landed object and heard voices from inside.

When he touched the side of the craft, his glove was partially melted. A door in the side of the craft opened briefly, then closed, and a blast of hot gasses from a grid-like port burned his clothes and chest. He also smelled a strong odor, often described as sulfurous or resembling rotten eggs, which lingered on his body for days afterward. Shortly thereafter, the craft took off, and Michalak fled the area. On the highway, sick and disoriented, he encountered an officer of the Royal Canadian Mounted Police.

The RCMP investigated the case but never drew any firm conclusions as to what sort of encounter Michalak had endured. He was hospitalized as a result of the physical trauma of the event, and soil samples later taken from the area by the RCMP were found to be "highly radioactive." According to the

RCMP report of 26 June 1967, the Department of Health and Welfare expressed concern that "others may be exposed, if travel in area not restricted."

Debunkers have attempted to cast doubt on Michalak's story by referring to conflicting accounts of what he was served in a hotel the night before the incident--coffee or beer?--and by highlighting the difficulties he later had, after his hospital stay, in finding the landing site again. The Condon Report, the study done under the auspices of the University of Colorado, at the request of the US Air Force, and published in 1968, looked into the case. Many feel that the Condon Report was very premature in its conclusions that UFOs were no longer in need of investigation, and that the purpose of the report may have been to mislead from the beginning. In the Michalak case, the Condon investigator seemed to think it significant that nothing unusual was reported from a fire watchtower a few miles away. Whether the fire watch personnel might have been less than vigilant, or whether they might have been reluctant to make a report that they felt could damage their credibility, does not seem to have been established.

The strong radiation reported by the RCMP, as well as the burn marks on Michalak's chest, however, which were photographed and well documented and left scars that remained for the rest of Michalak's life, are what make this case something other than a routine, interesting but unsubstantiated tale and make it instead a case worthy of note. The images of the burns Michalak sustained have been widely circulated in the age of the Internet, and they show a regular, square pattern of circular marks reminiscent of a chess or checkers board, but with round marks rather than rectangles. None of the accusations against Michalak--that he may have been served alcohol the night before, or his difficulty finding the site again after a hospital stay, explain the burns. No debunker has ever successfully shown what might have caused the burns, other than to leave the implication that, if the whole thing were a hoax, the marks would have to have been self-inflicted. Did Michalak do this to himself, in his desperation to make his story believable? That seems most unlikely, but those who claim that the whole story is a hoax should at least acknowledge that, if it were shown to be a hoax, Stefan Michalak would have to have thrown himself wholeheartedly into the enterprise.

My best judgment is that Michalak did indeed have a close encounter with a strange machine, perhaps an experimental military vehicle of some sort. The possibility of an extraterrestrial origin seems as impossible to prove as to disprove, so we are left to apply Occam's razor and speculate about which origin is more likely--terrestrial or extraterrestrial. Any conclusions that go beyond the empirical evidence will be merely speculative.

U stands for *Unidentified*.

Sources

Bernhardt, Darren. "Falcon Lake Incident Is Canada's 'Best-Documented UFO Case,' even 50 Years Later." http://www.cbc.ca/news/canada/manitoba/falcon-lake-incident-book-anniversary-1.4121639. Accessed 27 June 2018.

RCMP report of June 26, 1967. https://www.collectionscanada.gc.ca/ufo/002029-1300.01-e.html. Accessed 15 February 2018.

Stefan Michalak--Report of Unidentified Flying Object, Falcon Beach, Manitoba, 20 May 67. http://www.bac-lac.gc.ca/eng/discover/unusual/ufo/Documents/1967-06-26.pdf

The Scientific Study of Unidentified Flying Objects. 1968.
http://files.ncas.org/condon/
Commonly known as "The Condon Report."

THE CASH-LANDRUM INCIDENT

The story of what happened to Betty Cash, Vickie Landrum, and Landrum's seven-year-old grandson Colby around 9 p.m. on December 29, 1980 has been widely reported. The events of that night and the resulting ordeal, which lasted for years, are well documented and detailed by John F. Schuessler in his excellent book *The Cash-Landrum UFO Incident*. The evidence includes not only the account given by the adult witnesses, Betty Cash and Vickie Landrum, and Vickie's grandson Colby, but the bizarre medical symptoms, including burns and hair loss, suffered by all three witnesses in the aftermath. These symptoms puzzled physicians and were thought to be the probable result of some sort of radiation.

Though Betty Cash died in 1998 and Vickie Landrum in 2007, Colby Landrum has been interviewed and has discussed the event for public broadcast and podcast. He seems very open, coherent, and confident in his recollections, and makes no claims of extraterrestrial visitation. Some so-called skeptics claim or imply that the two respectable women, Betty Cash and Vickie Landrum, invented the whole story and somehow seriously injured themselves and the child to make the story more plausible, perhaps in the futile hope of winning a large award in a suit against the U.S. military. The notion that such a sham could have transpired without Colby Landrum ever having realized it was all a hoax is not credible. Something strange passed over the car on the Texas highway that night and changed three lives forever.

Briefly, the story is that, while driving home from a restaurant a few miles north of Houston, Texas, the three witnesses, Betty Cash and Vickie and Colby Landrum noticed a brilliant light over the road. The light was frightening, clearly like nothing they had ever seen, and was not far above the treetops. Colby Landrum has said the object appeared diamond-shaped, though it seems unclear if a solid diamond-shaped was present, or if the diamond shape were an appearance generated by the brilliance of the light. Flames were seen emanating from the lower portion of the object. Claims have also surfaced that the trees in the area were later found to be dead or scorched, and that the blacktop was blackened by heat before repaving. The claims of scorched trees and pavement seem hard or impossible to verify, though Schuessler did note that the road was in poor repair. Whether the repaving that was done not long after the incident is thought to have been normal maintenance or an attempted cover-up will likely depend largely on one's general attitude toward UFO phenomena.

Betty, who was driving that night, brought her car to a stop. The inside of the car began to grow hot from the energy being thrown off by the object, causing the plastic on the interior to soften. When the three passengers emerged from the car to look at the object, they found that the exterior of the car was becoming painfully hot as well. Vickie and Colby returned to the interior of the vehicle, while Betty remained transfixed outside for a few moments longer. However, she was soon driven back inside the car by the heat. The air conditioning was turned on, though previously the heat had been on, as the December night was cool. The dashboard of Betty's car was melted sufficiently for the mark of Vickie's hand to remain imprinted there.

The object was also accompanied by around two-dozen Chinook-type helicopters which, according to Betty, bore U.S. Air Force insignia. Even if the insignia had not been visible, it is difficult to imagine a non-military organization with access to a fleet of such aircraft. Chinooks are essentially flying buses--the two large overhead rotors that operate on the same plane make it impossible to confuse them with smaller military or civilian helicopters that have one large, overhead, horizontal rotor and a smaller rear rotor mounted in an orientation perpendicular to that of the large rotor. However, the US military has steadfastly denied any knowledge or involvement with the incident, leading many to reasonably suspect something was being covered up.

All three of the passengers suffered ill effects from exposure to the heat and whatever sort of radiation might have been present, but Betty, who had been outside the longest, suffered the worst effects. She was hospitalized on January 2, 1981, in Houston, and she remained there for over two weeks. Debunkers like to proclaim that "Betty's medical records were never released," but of course, this is either intentional falsehood or extremely poor research, as Schuessler's excellent book *The Cash-Landrum UFO Incident* includes various verbatim reports of Betty's examination by physicians during her hospital stay. Her symptoms included not only burns, blisters, and sores on her skin but also hair loss. As in the case of Stefan Michalak, it is difficult to credit the notion that such symptoms were self-induced in order to perpetrate a hoax. Some have even gone so far as to suggest she suffered from the rare mental disorder known as Munchausen syndrome, or factitious disorder, a well-known but uncommon problem identified in the 1950s. The essence of the disorder is that the patient pretends to be sick or self-induces illness in order to gain medical attention. None of the physicians who examined Betty, nor Schuessler, who spent much time with Betty and her friends and family, ever seemed to suspect any such bizarre explanation for her problems.

Other debunking claims have included the assertion that no one else saw anything unusual that night, but a dozen or so of the sort of helicopters the Air Force denied flying were also seen by the police detective Lamar Walker. Though the military claimed to have no involvement in the incident, the Air Force officer in charge of investigating Betty's and Vickie's claims, Lt. Colonel George Sarran, said he found them to be credible witnesses.

So what were Betty Cash and Vickie and Colby Landrum exposed to? The most likely explanation seems to me to be some sort of military technology, perhaps an experimental aircraft or flare gone badly wrong. The presence of the fleet of helicopters strongly suggests military involvement on some level, despite official denials. If the military concealed facts from the public in the Cash-Landrum case, it certainly would not be the first time secrets were kept. It is impossible to rule out the remote possibility of an extraterrestrial spacecraft--how can one prove a negative?--but nothing about the incident seems to require an alien explanation. Thus, a more mundane origin for whatever was in the Texas sky that winter night seems more likely. The one organization that may have the missing answers is the U.S. military. They do not seem likely to make any revelations soon.

U stands for *Unidentified*.

Sources

"Cash-Landrum Case Report 3-4-81 Redacted.pdf." https://app.box.com/s/wehhv92v81bcps8f35nqlq33yvwihf7y. Accessed 4 July 2018.

Collins, Curt. "More Details from UFO Witness Colby Landrum." http://www.blueblurrylines.com/2013/12/more-details-from-ufo-witness-colby.html. Accessed 5 July 2018.

Cox, Billy. "Cash-Landrum, Gone With the Wind." http://devoid.blogs.heraldtribune.com/11457/cash-landrum-gone-with-the-wind/ Accessed 4 July, 2018.

Collins, Curtis L. "From Their Own Lips: Betty, Colby and Vickie Tell Their Story." http://www.blueblurrylines.com/2013/07/from-their-own-lips-betty-cash-colby.html. Accessed 6 July 2018.

Hanks, Micah. "The Cash-Landrum Incident: A Case For Critical Review? – Micah Hanks Reports." https://jimharold.com/the-cash-landrum-incident-a-case-for-critical-review-micah-hanks-reports/. Accessed 7 July 2018.

"Rare Live Interview with Colby Landrum." (https://www.youtube.com/watch?v=pguNgh2F924). Accessed 18 February 2018.

Schuessler, John F. (1998). *The Cash-Landrum UFO Incident*. Houston: John F. Schuessler.

"Vickie Landrum's Phone Call to Report a UFO Encounter." http://www.blueblurrylines.com/2013/11/vickie-landrums-phone-call-to-report.html. Accessed 6 July 2018.

SOMETHING STRANGE IN THE DESERT: WHAT LONNIE ZAMORA SAW

Lonnie Zamora was a good cop, so when he saw a speeding car driving recklessly through Socorro, New Mexico, the town whose streets and citizens it was his duty to protect, he went after the speeder. No doubt, he little suspected to catch something far stranger, which would not only make history but earn him unwanted notoriety.

These events took place on the evening of April 24, 1964, before the time when humans would walk on the Moon or land spacecraft on Mars to send back pictures of the barren-looking surface. The early 1960s were in many ways the end of an age of innocence--at least when "innocence" is considered in its secondary meaning, as lack of knowledge. Months later in the same year as Lonnie Zamora's sighting, Mariner 4 would, on November 28, 1964, be launched on its way to Mars, where it would go into orbit in the summer of 1965. The Soviets had launched Sputnik 1 less than seven years earlier, and Telstar, the U.S.'s first communications satellite, had been launched less than two years previously. So in 1964, one could, without completely ignoring well-established scientific fact, still reasonably speculate that other planets in the Solar System might, just might, be home to civilizations that were capable of sending spacecraft to Earth.

If there were life on Mars, why had nothing like radio or television signals been received from the red planet? As early as 1901, the engineering pioneer Nikola Tesla reported his belief that he had received a radio transmission of extraterrestrial origin. By the 1960s, however, no confirmed intelligent messages from Mars had been announced, which must have been disappointing to those who still clung to the slender hope of finding intelligent life elsewhere in the Solar System. Astronomers and astrophysicists already had strong evidence to support the belief that Venus was far too hot for life as we know it, not to mention other drawbacks based on composition of the atmosphere. Other planets, Mercury, or the gas giants--Jupiter, Saturn, and beyond, were unlikely candidates for life of any sort--Mercury would be too blisteringly hot and could have no atmosphere, and the gas giants, such as Saturn, seemed unlikely places for ET to call home as well. Still, despite the lack of evidence of life elsewhere in the Solar System, that possibility had not been totally disproven. Thus, speculation that ET might visit from somewhere closer than the nearest star besides the Sun did not automatically confine one to the lunatic fringe.

Such matters were no doubt far from Lonnie Zamora's mind when he was attempting to deal with the speeding car that April evening in 1964. Presumably the speeder got away without even a warning, because, on a little road a couple of miles west of Socorro, Zamora heard a loud, disturbing noise from somewhere off the road. He stopped to investigate. In the distance, he saw what he thought might be an overturned car, possibly with passengers who would need help, but on coming closer he spotted an "egg-shaped" object roughly the size of an automobile. Zamora also said he saw two figures in "white coveralls," near the egg-shaped craft, which he later refused to say were human beings. He referred to at least one of them as a "white object" that, he thought, "turned and saw" him. Again, there is room for endless speculation about whom or what he may have seen, but no firm answers. Realizing the situation demanded the attention of additional officers, Zamora called the state police for assistance in dealing with whatever the object might be. However, the craft soon flew away, with a frightening expulsion of flames from beneath.

The evidence that something strange was present in the desert and sky that day consists of Zamora's word and the word of other witnesses who saw or heard a strange object or unusually loud noises. Physical evidence at the site was also found, as discussed by Stanford and others. Both Zamora and the New Mexico State Police officers who responded saw burning vegetation and remains of burned vegetation near the marks left from the "legs," or struts or landing gear, on which the craft rested. The imprints and evidence of burning was also later observed by various other witnesses.

Unidentifiable organic substances and clumps of fused or melted sand, somewhat similar to lava, were reported to have been found where the object had been seen.

However, based on Druffel's discussion of the incident, the samples of fused sand and strange organic material appear hard to confirm. Like the alleged spaceship and rumored alien bodies supposedly found near Roswell, they seem to have vanished into the vaults of the Pentagon or a secret hangar in Areas 51, never to be seen again by civilian eyes. As with Roswell, the lack of access to materials does not mean they did not exist, but it is hard to base any firm conclusions on materials that cannot be properly examined by qualified scientists and engineers, whose results are then verified by others. The key difference here is that plenty of other evidence documents what Lonnie Zamora saw.

When Walter Shrode, of radio station KSRC, interviewed Zamora later in April, 1964,

Shrode mentioned that an Albuquerque television station had reported a UFO fifteen or twenty minutes before Zamora witnessed the object near Socorro. If indeed these two sightings were of the same craft, it would have had to be travelling at around three hundred miles per hour to cover the 75 miles from Albuquerque to Socorro in just fifteen to twenty minutes. Zamora also told Shrode that he did not believe the craft was an object from outer space. This statement would lead one to conclude that Zamora did not believe he had seen aliens there, either.

In the same interview with Shrode, Zamora said "they," presumably referring to government or military officials, had instructed him not to discuss any marks or insignia on the craft. The question of marks or symbols on the craft, of course, raises another point that can give rise to endless speculation. Perhaps the officials knew what the insignia meant and did not want such knowledge to become widespread, or they did not know and wanted to conceal their ignorance, which would be exposed if the insignia were widely discussed.

Few who have ever looked into the incident have accused Lonnie Zamora of fabricating the whole story or deliberately perpetrating a hoax. No credible case can be made that Zamora told anything but the truth of what he saw and heard. By all accounts, he was simply not the sort of person who would make up such a story. Furthermore, there was too much evidence on the ground and too many other witnesses to doubt that something had happened. So the unanswered question is, what landed in the desert that evening? Some say is was indeed a hoax, but one perpetrated successfully by high school or college students, to trick Zamora. The alleged motive of the never-identified--that is to say, fictional--students, was that Zamora had supposedly been too strict in his duties as a policeman.

The revenge prank-hoax theory, aside from having no reliable first-hand witnesses, simply makes no sense. Its advocates assert that the alleged pranksters were unwilling to come forward once they saw how seriously their little joke was being taken by the federal government and the military, but this explanation, of course, assumes that the anonymous hoaxers actually existed. Moreover, who in his or her right mind thinks of an elaborate scheme such as building a fake UFO as a way to take petty revenge? And why would such a scheme even seem vengeful? No reasonable person would expect a person of integrity such as Lonnie Zamora to be harmed by uncovering a UFO hoax. Letting air out of car tires, putting the prank-victim's name on unwanted magazine subscriptions, sending unordered pizza deliveries at 11:30 p.m., egging the windows or toilet-papering the trees at the victim's house--these are effective pranks. But building a fake UFO in the hopes that one can somehow, through an act of reckless driving, trick the victim into witnessing and reporting it--this hardly comes to any sane person's mind as an effective revenge hoax.

But, supposing such a bizarre plan had been conceived by students, would it not have occurred to them that the ticket and fine they would almost surely have received, and perhaps also an afternoon or night in jail as well, on charges of endangering public safety, were sufficient reasons to hatch a different plan for revenge? If the alleged--*fictional* is a better term to describe them--hoaxers were bright enough to build a fake UFO as convincing as what Lonnie Zamora saw, they would also be clever enough to come up with a better plot to seek vengeance.

Listening to Zamora's discussion of the object, one has to either accuse of him of cowardice and exaggeration, as he said a frightening volume of flame emerged from the craft as it took off, or admit that, whatever else the craft may have been, it was the product of sophisticated engineering. It could not have been an amateur contraption made of candles and ordinary balloons, as some have suggested. The marks and evidence left on the ground also indicate that a complex device was at work. If students had somehow managed to acquire not only the materials but the skills to build such a craft, they should have been working for NASA, not attempting pranks on local policemen.

So what did Lonnie Zamora and others see and hear? Various possibilities have been put forward, including the usual suspects--balloons, ball lightning, dust devils, and some prototype of a Lunar lander or module being developed for one of NASA's Moon missions. These explanations all fall short. The object did not closely enough resemble any of NASA's Lunar craft, which were not being flown in that area then, anyway, and balloons do not scorch sand and set vegetation on fire or fly as rapidly as the strange object Zamora saw. The object simply did not look anything like dust or lightning, unless Zamora was fabricating a wild story just to gain attention. All accounts suggest Zamora actually disliked the attention. When he eventually left police work, he tended to avoid further questions about the sighting. In any case, dust devils do not scorch the soil, and lighting does not leave imprints from landing gear.

The Air Force and the FBI took an interest in the case, and J. Allen Hynek, of Project Blue Book and *Close Encounters* fame investigated, without much success, as far as any public announcements ever indicated. If anyone in the government or military establishment knew what the craft was, they must have been very successful in concealing their knowledge from other official investigators. Such secrecy does not seem all that unlikely--professional secret-keepers are no doubt often very good at what they do. Only the inept ones make the news.

The alien-spacecraft hypothesis cannot be disproven, though it seems more likely that a secret, never-to-be-revealed experimental aircraft was being tested. The exhaust-like flames and noise and rapid motion suggest the craft was rocket-propelled. It it were of extraterrestrial origin, one would have to suppose that it was some sort of landing craft sent out by a larger ship. Not only would a vehicle the size of a car seem rather cramped for a long voyage through the lonely reaches of interstellar space, but any rocket engines we can conceive would be inadequate to power interstellar travel. In the early 1960s, of course, it was reasonable to suspect Martian origins, but not today. The craft was probably of terrestrial but unexplained origin. But until we have conclusive evidence, which seems less likely as the decades pass, the case remains open for speculation.

U stands for *Unidentified.*

Sources

Druffel, Ann. *Firestorm: Dr. James E. McDonald's Fight for UFO Science.* Granite Publishing, 2003.

"First Space Missions: Mariners to Mars." https://www.jpl.nasa.gov/jplhistory/missio

n/mars-t.php. Accessed 11 July 18.

Hynek, J. Allen. *The UFO Experience: A Scientific Inquiry.* New York: Ballantine, 1972.

"Lonnie Zamora interview, April 25, 1964." (Audio file). http://www.noufors.com/audio/Cops%20and%20Saucers/07.mp3. Accessed 14 July 2018.

"Lonnie Zamora interview, April 25, 1964." (Transcript). http://www.roswellproof.com/Socorro/Socorro_Zamora_interview.html. Accessed 14 July 2018.

"Officer Lonnie Zamora is interviewed on the phone by NICAP's Streeter Stuart, April 29,

1964." http://www.roswellproof.com/Socorro/Socorro_Zamora_interview.html. Accessed 14 July 2018.

Stanford, Ray. *Socorro "Saucer" in a Pentagon Pantry*. Austin, TX: Blueapple Books, 1976.

Tesla, Nikola. "Talking with the Planets." 9 Feb 1901. *Collier's Weekly* p. 4-5. http://teslacollection.com/tesla_articles/1901/colliers/nikola_tesla/talking_with_the_planets. Accessed 11 July 2018.

"The Fused Sand." http://www.educatinghumanity.com/2011/06/ufo-ufo-police-witness-officer.html. Accessed 12 July 2018.

THE UFO LANDING IN TRANS-EN-PROVENCE, FRANCE

One very convincing case that has not been very much discussed in the United States, at least in comparison to the misguided sensationalism devoted to Roswell, is the Trans-en-Provence incident. Not only did a credible witness see a strange object land, but clear, difficult-to-explain traces were left on the ground. Those who try to infer some fragile connection between UFO sightings around the world will point out that the sighting took place on January 8, 1981, just a few days after the Cash-Landrum incident and the Rendlesham Forest events.

Renato Nicolai is variously described as a farmer, technician, contractor, and retiree. Whatever his primary occupation, reports agree that he claimed to have seen a convex, circular object, five or six feet in height and width, of unknown origin descend while he was working outside near his home in Trans-en-Provence in the south of France. After landing for a brief period, the object flew away, leaving round imprints on the ground. Unlike some who witness UFOs, Nicolai was unharmed and apparently not particularly frightened, nor was he anxious to gain notoriety from the incident, as he did not report the sighting to the police until the next day, on the urging of others. Also unlike others, as in the Michalak and Cash-Landrum incidents, Nicolai did not experience illness or physical symptoms in the wake of the encounter.

Analysis by the French space agency, GEPAN, found that the imprints left in the soil by the object would have required a weight or impact pressing down with the force of several tons. Laboratory reports showed definite changes to the soil that indicated not only great pressure but also heating to a temperature of several hundred degrees. So it seems that, if indeed this incident were an unlikely, carefully staged hoax, it would deserve an award alongside the Socorro and Cash-Landrum and Michalak cases for elaborate preparation and attention to detail.

Debunkers have attempted to dismiss the case by suggesting that the marks might have been made by an ordinary object, such as a tractor, even though the soil was clearly subject to temperatures no tractor would produce. Debunkers also question the soil analysis because of unspecified deviations from scientific methods in gathering plant and soil samples, but what these alleged mistakes were or how they would have negated the results is never made clear. Moreover, the samples later collected and tested by Vallee and others were consistent with the original analysis that showed unexplained changes to soil in the landing site. At this point, the debunkers' claim seems far weaker than the assertion that something strange was seen that evening.

Assuming Nicolai saw exactly what he reported, what could it have been? An experimental terrestrial aircraft of some sort seems possible, though the object does not seem to have been large enough to accommodate more than, at most, one person, presumably a pilot. Was it a remotely operated craft of some sort? No government agency, military organization, or private corporation has ever claimed responsibility for the object. As with the other truly mysterious cases, we cannot say for certain what it was, and therefore the possibility of an extraterrestrial origin, unlikely as it may seem, cannot be ruled out with absolute certainty.

U stands for *Unidentified*.

Sources

"France Opens Up Its UFO Files." 22 March 2007. https://www.newscientist.com/article/dn11443-france-opens-up-its-ufo-files/.

Accessed 18 July 2018.

Klass, Philip J. *The Skeptics UFO Newsletter.* July 1 1997. https://www.csicop.org/specialarticles/show/klass_files_volume_46. Accessed 18 July 2018.

Sturrock, Peter A. *The UFO Enigma.* New York: Warner 1999.

UFO Case 4: The Trans-en-Provence Case. https://sites.google.com/view/ufoskepticorg/ufo-cases/ufo-case-4-the-trans-en-provence-case. Accessed 18 July 2018.

Vallee, Jacques F. "Return to Trans-en-Provence." http://www.jacquesvallee.net/bookdocs/Return_to_Trans-en-Provence.pdf. Accessed

25 July 2018.

Velasco, Jean-Jacques. "Report on the Analysis of Anomalous Physical Traces: The 1981 Trans-en-Provence UFO Case." https://www.scientificexploration.org/docs/4/jse_04_1_velasco.pdf. Accessed 25 July 2018.

Wilson, Jim. "When UFOs Land." *Popular Mechanics,* May 2001, p.64-7. https://sites.google.com/view/ufoskepticorg/ufo-cases/ufo-case-4-the-trans-en-provence-case. Accessed 18 July 2018.

Roswell, or How a Lot Was Made of Very Little

The Roswell incident has an irresistible appeal. Like others with an interest in UFOs, I have read account after account, many of them attempting to prove an alien spacecraft crashed in New Mexico in 1947. I keep hoping someone can convince me, frightening as such a conviction might be. If aliens had been around as early as the 1940s, what could their agenda be? (Alternately, one could ask, what might aliens do to us that we do not seem anxious and willing to do to ourselves?) Yet, Mulder-like, I want to believe.

But the evidence suggests little besides the crash and recovery of a piece of classified US military equipment. Roswell is a good story. It's great folklore. Roswell has been the subject of numerous rhetorical flourishes by many talented writers. But Roswell is not proof of ET. In fact, extraterrestrial involvement at Roswell is so poorly supported that the whole incident was an early example of Warholian "fifteen minutes of fame," and then it was mostly overlooked until the 1980s.

If the evidence of alien involvement is unconvincing, how and why was the case eventually made so famous that it has become synonymous with UFO sightings in the popular lexicon? A three-fold answer can be given. First, by the Air Force's own admission, something did crash or land in the New Mexico desert in 1947. There was a real event, but most likely not the one the true believers want to imagine. Second, there was probably a real attempt to quell public discussion of the event by the military--which hardly proves alien involvement. The military has kept many secrets having nothing to do with extraterrestrials. And third, by the 1980s, there was money to be made. UFOs have become big business in some corners of the world, including the tourist trade; book, TV, and movie deals; and speaking fees at UFO conferences. The Roswell case was tailor-made to be exploited in all such areas.

The cover-up seems to have begun almost immediately. The now famous July 8, 1947 edition of the *Roswell Daily Record* proclaimed in a bold front-page headline,

"RAAF captures Flying Saucer in Roswell Region." We will probably never know who first uttered the word "alien" in connection with the object, probably a balloon carrying classified technology. There is room for speculation about what exactly came down, if one mistrusts the Air Force claim from the 1990s that it was a Mogul balloon. A top-secret aircraft, possibly even of Soviet or other foreign origin? Perhaps the military, in an effort to distract attention from devices it did not want publicized, decided to refer to the material as a flying saucer or disk, or perhaps the military heard rumors buzzing around Roswell that a saucer had crashed and simply decided to encourage the misperception. In either case, they must soon have realized that a flying saucer as the cover story was going to draw more attention, rather than having the desired effect of tamping down public interest. Thus, on July 9, the *Daily Record* reported that no, the object was not a disk, and it had only been a balloon. So nobody paid much attention to the case for decades.

By the 1980s, the 1977 movie *Close Encounters* had done much to bolster the attention paid to UFOs in the popular culture, and the continuing fascination with space-themed entertainment, from television to the big screen--*Star Trek, Star Wars, 2001: A Space Odyssey* and its follow-up *2010: The Year We make Contact*--all proved that, whether presented as fact or fiction, aliens and extraterrestrials made good entertainment. Human success in space--including Moon landings and the Space Shuttle program, not to mention the Voyager and other spacecraft exploring the far reaches of the Solar system, kept the idea of space travel alive and well in the public mind. Thus, there was money to be made. After a few books and TV shows focused on Roswell, it was well on its way to becoming famous as the place where the aliens crashed.

If an alien spacecraft did crash near Roswell, a lot of important people who should have known were somehow kept in the dark about it. Declassified documents now show that Air Force officers believed no UFO had been recovered, well after the alleged Roswell crash. Those who believed no UFO had been recovered included Col. Howard McCoy and the often-cited General Nathan Twining, beloved by students of UFO lore for his insistence that the UFO phenomena were real and must be investigated. It is hard to believe the Air Force covered up a retrieved alien spacecraft and alien bodies so successfully that even the people one would expect to be in charge of any cover-ups did not know about them.

Roswell has few credible first-hand witnesses, and the information supplied by the credible witnesses does not support the alien hypothesis. Advocates of the alien hypothesis rely heavily on second- or third-hand accounts from friends or relatives, including anonymous sources. Often these accounts include stories of threats not to talk about what had been seen. The unlikely implication we are encouraged to draw from these stories of intimidation is that the military was hiding aliens, rather than keeping classified but ordinary technology from entering the public record.

Conversely, there is nothing surprising about the Air Force telling people not to discuss the relatively mundane but secret technology that probably did land or crash.

Jesse Marcel, Jr., is widely reported to have seen what was found by his father, the Air Force officer credited with bringing the alleged flying saucer wreckage back to the Roswell base. Marcel, Jr., saw the balloon wreckage for a few minutes when he was a sleepy kid--that's hardly proof of ET. No one wishes to impugn the Marcels' integrity, father or son, both of who seem to have been good, honest, decent people. But neither of them were aerospace engineers, nor did they perform qualitative analyses of the wreckage in a chemical laboratory. They saw some material they did not recognize and apparently jumped to the conclusion that it was alien. There is simply no convincing evidence that the materials--tough metals or plastics, some with marks the Marcels did not recognize (which have somehow now become known as "hieroglyphs")--were extraterrestrial.

What is convincingly reported can be readily attributed to mundane events. Many people reported seeing a bright light in the sky, easily explained as lightning or a meteor. Those who think a bright meteor around the time of the crash is too coincidental have not spent much time looking at the night sky--meteors are quite common. Alleged witnesses also reported seeing something on the ground outside Roswell which they could not identify. Many of them, such as the intoxicated trespasser who claimed to have watched the recovery of an alien craft while he and his girlfriend hid in the bushes, had no expertise in aviation technology or military equipment. Some alleged witnesses may really have seen something interesting, but their accounts are far from proof of alien contact.

Other accounts run along the lines of a "researcher" having interviewed John, who was told by Mary that her neighbor saw a UFO crash--and John has signed a "sworn notarized affidavit" that this story is exactly what he was told! Legal terminology and useless legal-style documents are good smokescreens in matters where there are no actual legal consequences for misremembering or even fabricating a story. Or a story is told by someone who at the last minute was intimidated by mysterious official threats, and withdrew his/her account or requested that his/her name not be used. Affidavits and anonymous witnesses who have allegedly been intimidated by the government or military are clever rhetorical devices, but not convincing evidence. It all makes for a good bit of folklore, but not substantial proof of an extraterrestrial crash. A lot of questionable witnesses do not make for an unquestionable story.

Yet, where there is smoke, there is fire. Something came down out of the sky, and the Army did not want anyone to know what it was. This part of the story is uncontroversial--the Air Force has since admitted as much. But the alleged evidence that what came down was extraterrestrial is lacking. The so-called experts on Roswell cannot seem even to agree on the location of the alleged crash, despite their insistence that they have interviewed people who saw dead aliens and/or hauled the crashed saucer away to a military lab or vault. The original 1947 Roswell local newspaper article was the best evidence that something unusual came down somewhere in the desert, and the USAF explanation given in the 1990s, that a secret device for monitoring Soviet nuclear activity was what really crashed, makes sense. If one rejects the Air Force's 1994 conclusion because, as some accounts claim, the debris field was too large or some of the material supposedly did not look (to non-experts) like a balloon, the most reasonable conclusion is still that some terrestrial craft crashed.

Then there are the claims that modern technology developed since the late 1940s and early 1950s was based on technology found in the craft from Roswell. This, despite the history of technology showing the clear path of human innovation and step-by-step development of everything from microwave ovens to cell phones, no alien inspiration required. It's possible that, if ET did come to Earth, he or she might have brought along a microwave oven to make popcorn, but there is no proof of this. Aliens probably don't like popcorn, anyway--the salt is bad for their silicon-based metabolism.

However unbelievable the claims that humans reverse-engineered an alien spacecraft, this aspect of the incident makes it fascinating--not as research into aliens but for the insight it provides into human nature and psychology. Why are some people so eager to tell such stories, and so many others so willing to credit them as fact? Along with stories of seven-day Biblical creationism 6,000 years ago, tales of the hollow (or flat) Earth, flash-frozen mammoths unearthed with fresh daisies in their mouths, or conspiracy theories that say the Apollo Moon landings were a hoax, the stranger the claim and the more insupportable it is, based on empirical evidence, the more appealing it is to some. We humans love our fictions, even when we do not know they are fiction.

Alien bodies at Roswell? Rumor and legend. People have claimed to have seen what they thought were alien bodies, or to have been told by anonymous sources that they, the anonymous ones, saw aliens. Yet it is strange that no alien tissue has ever been publicly presented for analysis. Did none of the alleged witnesses think to snag a hair or fingernail or claw or a bit of skin or protoplasm, whatever might be available on an alleged alien body? Then, too, the Army was supposedly shopping for body-bags or child-sized coffins, but for what purpose--to give the aliens a proper Christian burial? If the most significant discovery since the unearthing of dinosaurs or Neanderthal bones were made, would the Army have put them in coffins from the local undertaker and buried them to rot and be forgotten? It hardly seems likely.

If in fact there were dead bodies in circulation at the time, it seems more likely that they were the results of an ordinary crash or experiment gone wrong. Might they have been the bodies of chimps or monkeys used for some purpose the military did not want publicized? Occam's Razor suggests that is a more likely possibility than aliens, especially since the witnesses tend to say they saw the "aliens" for a moment from a distance, or in mangled condition.

Then there are the stories about military officers calling the local civilian undertaker for advice about how to handle a dead body before burial. Bear in mind, these events supposedly took place only a couple of years after World War II. One must be desperate to believe in order to accept the notion that, in the entire post-war U.S. Army, not a single person who had ever handled a dead body and knew what to do with one could be found. Unfortunately, after the carnage of war, such soldiers would have abounded. Many would have been discharged, but certainly not all. There would have been no need to ask revealing questions of loose-lipped civilians while working frantically to keep the secrets.

The military would almost certainly have preserved the alien bodies, if it had any, in vats of alcohol or formaldehyde, or flash-frozen as in the episode of *The X-Files* called "The Erlenmeyer Flask"--so a large purchase of such preservatives would be more significant than coffins or body bags. If someone comes up with an actual preserved alien body or piece of inexplicable technology more puzzling than microwave ovens or cell phones, that will be truly exciting. Until then, the most interesting aspects of Roswell lie in its cultural status as folklore.

Does such a conclusion mean all the

witnesses who claimed to have seen aliens were lying or hallucinating? No. But consider the alternatives: the military, in its zeal to protect the top-secret project of monitoring Soviet nuclear activity, fabricated, or at least encouraged, the crashed-alien-ship story as a smokescreen. If no dead chimps were available, and if the mannequins the Air Force had were unconvincing, perhaps some artificial alien bodies were created in a lab somewhere, or borrowed from the prop-shop that supplied legendary film-maker Ed Wood with materials. Thus, the witnesses saw what the "higher echelons" wanted them to see. The Air Force must have known that, if anyone ever dug deeply into the matter, enough wild tales and conflicting reports would be uncovered to keep the Soviets and the American public well confused. Such a scenario, while it may seem an unwieldy approach to national security, is at least as plausible as the alien-crash hypothesis.

But maybe aliens did crash in the New Mexico desert. I cannot prove they did not. I simply remain unconvinced that they did. While there is a lot of what some take as evidence, it does not support the profound conclusion that ET arrived, DOA or otherwise. Somewhere, someone is working on a book that promises amazing new revelations of aliens at Roswell. Maybe then we'll at least know where the object landed. I'll buy it, and I'll read it, in the slender hope of being convinced. If not convinced, maybe I will at least be entertained, and further evidence of the strange convolutions of the human psyche will accrue.

Recommended Reading

This discussion of the Roswell case is based on years of reading and interest in the subject, including all the big names, who shall here remain nameless, as the advocates of the alien-crash hypothesis seem like nice, sincere people, whose books, blog posts, and commentary I have enjoyed, without being convinced. The reader who is well-versed in UFO studies will be familiar with the names. For those who are not familiar with Roswell lore, here are some good places to continue the search:

Clarke, David. The UFO Files. Kew, Richmond, Surrey, UK: The National Archives, 2009.

Colavito, Jason. "Flash-Frozen Mammoths

and Their Buttercups: Yet Another Case of Repetition and Recycling of Bad Data." http://www.jasoncolavito.com/blog/flash-frozen-mammoths-and-their-buttercups-yet-another-case-of-repetition-and-recycling-of-bad-data. Accessed 4 August 2018. The mammoth story is another perfect example of how a lot is made of very little, and serves as a cautionary tale about drawing conclusions without tracing information back to physical evidence or primary sources.

Pflock, Karl T. *Roswell: Inconvenient Facts and the Will To Believe*. Amherst, New York: Prometheus Books, 2001.

The Roswell Report: Fact vs. Fiction in the New Mexico Desert. http://www.roswellfiles.com/Articles/AirFo

rceReport.htm. Accessed 2 August 2018.

Interstellar Travel for Fun and Profit, or, How To Build Your Own UFO

Could aliens actually have visited the Earth? This question is often answered with speculation about possible alien technological advances that would make interstellar travel easier than it now seems. But given out our current human technology, just how hard would it be for us to travel to other stars? Or would it be possible at all? The answer, I think, is that we could in fact send humans to other stars. It would be hard, and it would require immense resources and cooperation, but it could be done. I know of no convincing evidence that aliens have visited the Earth (or that they have not), but if we can develop interstellar capability, then it is conceivable that extraterrestrial species may also have done so, or will do so.

What are the challenges for interstellar travel? First, of course, the distances between stars are huge. Even the nearest stars, the Alpha Centauri system, are over four light years away. Thus, if the wildly optimistic human dream of traveling at even ten percent of the speed of light comes to fruition, the one-way trip to Alpha Centauri would take over forty years--considerably more than forty years, considering that a spaceship would not instantly accelerate to ten percent of the speed of light, nor would it instantly decelerate to orbital speed on reaching the planet. Just reaching full speed would likely require months or years, depending upon the method of propulsion. This means a round trip for a human adult is unrealistic, unless one is ready to consider the prospect of a space traveller making it back to Earth only when he or she is over a hundred years old: sending an astronaut who is less than twenty years old seems problematic for various reasons, and add eighty years for the round trip. This math is of course, optimistic and leaves no time for exploration of any planets, should even this nearest star system have planets worth exploring.

Nor is travel faster than a small percent of the speed of light realistic, given our technology or likely future technologies. Not only must serious propulsion and engineering problems be overcome, but the ship traveling at a significant fraction of the speed of light might shatter itself to bits when crashing into even a tiny piece of debris. A tiny meteoroid or fragment of a comet the size of a grain of sand might be fatal if it collided with a ship traveling at speeds sufficient to take people to other stars in a single lifetime. Consider a bullet fired from a gun as an example. Drop just the bullet on your toe, and it will not leave even a bruise, travelling at the slow speed given it only by the force of gravity between your hand and foot. The factor that makes it deadly is the velocity imparted by being fired from a gun. Velocities are relative, so it makes no difference whether we consider the meteor to be travelling at immense speed or the ship to be travelling at immense speed. The key factor is the speed at which the objects collide. The faster the ship travels, the more likely a fatal impact becomes.

Robert Goddard, famous for developing rockets that used liquid fuel in the early twentieth century, also speculated about the possibility of interstellar space flight. He harbored no illusions about the difficulty and risk of such ventures, and so suggested that several ships should be sent in order to increase the chances that at least one might reach another star with living humans capable of exploration or colonization.

The conclusion, then, is that a single human lifespan will likely never be long enough to travel to another inhabited or habitable planet outside the Solar system. Well, if it were easy, someone would already have done it.

However, it might yet be possible for humans to arrive on a planet orbiting another star. Robert Goddard was also one of the first to suggest ways to overcome the problems of limited human lifespan in interstellar travel. Goddard suggested that the crew of a ship might be put into a "granular state" to endure the cold of space, with the pilot being awakened periodically to keep the ship on course.

This solution sounds like what we would now call cryogenic sleep or suspended animation, a useful plot device for writers and movies such as *Alien* and *Planet of the Apes*. Fans of the original *Star Trek* series will recall that in the episode entitled "Space Seed," Khan Noonien Singh and his companions are awakened from suspended animation, having traveled in space for hundreds of years. But putting a healthy human being into some sort of long-term stasis to survive centuries of space flight is still firmly in the realm of speculative fiction. It is not a technique we can now use, nor one we can assume aliens will use to visit us.

Konstantin Tsiolkovsky also speculated that interstellar travel might be possible using a more promising method, the generation ship. According to this plan, generations of travelers would live, reproduce, and die on the way to other stars, leaving their offspring to carry on the voyage. So far, of course, no humans have ever attempted such a feat, outside the realm of science fiction. However, in principle, this approach seems the most promising for human interstellar travel. Thus, assuming no extraterrestrial species has either a technology far advanced beyond our own, nor a tremendously longer life span, generation ships are the most likely method whereby aliens might have visited the Earth.

What are the challenges in such an enterprise? First, the travelers, terrestrial or otherwise, would need to be very determined, whether motivated by curiosity and a desire to explore, the desire to leave a dying planet in search of another home, a desire simply to take part in something larger than themselves, or perhaps some alien motivation humans cannot comprehend. The first generation would expect to die in space, without ever setting foot (or tentacle) on another world; that privilege would be reserved for distant offspring, and then only if the intervening generations did everything right.

The building of such a ship by humans would also be a monumental undertaking. The ship would be far too large to lift off the surface of the Earth using even the largest rockets, so it would have to be assembled in orbit. International cooperation would be required, perhaps over several decades. As there is no guarantee that the ship will reach another star, several ships would need to be sent in order to increase the odds of at least one reaching its initial destination. The United States and NASA justifiably congratulated themselves on putting humans on the Moon in less than a decade, but that project was a walk in the park in contrast to an interstellar generation ship.

The generation ship would face not only external but also internal hazards. Imagine that you are a child or grandchild of the ship's original occupants. How would you feel about your ancestors' decision to make this voyage? Would you remember them as bold pioneers and eagerly embrace the challenge of carrying on their legacy? Or would you resent being doomed to live and die in a closed environment, never to feel the Earth's wind on your face, smell a fresh morning breeze off the ocean, or watch a sunset on a Pacific beach? All the while, of course, most generations would know they would never set foot on an extraterrestrial planet. What if the second, or third, or fiftieth generation decided to turn back? Would you lead a rebellion to turn the ship back to Earth, or join the rebellion of someone else started one? The interstellar ship is often compared to Noah's ark, but perhaps a more appropriate and realistic Biblical metaphor is the Israelites' forty years of wandering in the desert--if the figure of forty years is multiplied exponentially.

Assuming the descendants of the first generation remain committed to the project their ancestors undertook, the social challenges are nevertheless formidable. Successive generations would need to be trained in every aspect of the ship and its operation, from biology to agriculture and various engineering specialties; mechanical engineers, electrical engineers, and experts in propulsion would all be essential. Medical doctors and healthcare workers--both mental and physical--would be equally important. Educating the next generation would be essential to survival. The ship would be an isolated society, so any breakdown in the social fabric could cause the entire project to fail. If any of the descendants ever returned to Earth, the psychology and sociology of the ship's inhabitants would likely provide material to keep entire university departments busy for decades.

The difficulties of maintaining a relatively small, isolated society should not be underestimated, as witness the problems of the Biosphere project in the Arizona desert in the early 1990s. Aside from problems of engineering and the basics of biology and survival--it is hard to maintain food and oxygen supplies with limited resources--it is said that the participants in the experiment did not necessarily get along personally. If humans set out on a generations-long voyage to other stars, who is to say that, even if the first generation carries out their part of the trip, as planned before departure from Earth orbit, the second or third or fourth, etc, generations will know or care what their grandparents had in mind? Science fiction gives us examples of how a generations-long voyage might go awry. For instance, Frank Robinson's *The Dark Between the Stars*, *Ship of Fools* by Richard Paul Russo, and Robert Heinleins' "Universe" and "Common Sense" all explore the human aspects of interstellar flight. Even if the technology holds up, will the ship's inhabitants a dozen generations later even still understand how to operate it or what its purpose is? Will they still necessarily even understand that their immense ship is tiny in comparison to planets and stars?

At some point in its voyage, the ship will be so far from Earth that messages sent between Earth and the ship, even at the speed of light, will take a frustratingly long time, and the ship will become essentially an isolated civilization, perhaps doomed to go the way of the inhabitants of Easter Island or the first settlers at Roanoke, their fate a mystery.

Other possibilities for interstellar travel exist, of course. One can hope that purely speculative technologies, based on physical principles we have no current ability to comprehend, might someday make travel at the speed of light or beyond possible, or that we might somehow find a way to use wormholes to jump across immense distances. Argument by analogy cautions us not to be arrogant about our current physics, as our current technology would seem like magic to someone born in the 1700s. Astonishing inventions may await. Yet conclusions based on by analogy are uncertain at best. Any new physics--for instance, answers to questions of dark matter and dark energy--may do nothing to make interstellar travel easier.

Yet another possibility, and one that is currently quite realistic, is of course, is the use of robotic spacecraft with no biological life aboard. We have already launched what will eventually be an interstellar spacecraft, as Voyager 1 left the Solar System in 2013. However, it will not become a truly interstellar craft, reaching the vicinity of another star, for tens of thousands of years. Moreover, it is essentially little more than a flying billboard, carrying the messages--a phonograph record and images--that may or may not prove meaningful to any extraterrestrial discoverers. We will almost certainly never know its fate.

But with advances in artificial intelligence, terrestrial robotic spacecraft may be launched on interstellar journeys and programmed to report back to Earth, centuries or millennia from their launch date. (It is worth asking, will anyone be here to get the message? Will anyone recognize the message when it arrives?) Similarly, if we can launch robotic craft, maybe ET can, too. Extraterrestrial intelligence, if it exists, might use artificial intelligence to explore the galaxy. These possibilities are examined by means of speculative fiction in the final chapter of this book.

Another important question is, what are extraterrestrial physiology and lifespan like? We have no empirical data to point to an answer. Or rather, we have data regarding only terrestrial physiology and lifespan. However, if we extrapolate from this limited set of data, we can conclude that biology is spectacularly diverse in many respects, including lifespan. The oldest known tortoise has lived over a hundred and eighty years. Other species can put themselves into states of hibernation or dormancy, and plant seeds thousands of years old have proven capable of germination. Sponges and trees also live for thousands of years. Lifespans vary widely among mammals, from a year or two for mice, to many decades for other species. We can conclude from this diversity that mother nature puts no hard limits on lifespans, and there is little correlation between the the complexity or intelligence of an organism and its lifespan. We are more intelligent than trees, yet they may outlive us. We are more intelligent than mice, and yet we outlive them. So it is plausible to imagine intelligent extraterrestrial beings who not only live for thousands of years, but who might be capable of naturally extending their lives indefinitely through hibernation. In other words, they might reap the benefits of stasis or cryogenic sleep--a technological science-fiction dream for humans--without needing any special technology. Like our robotic spacecraft, alien life might plod slowly between the stars for millennia and arrive on Earth ready to explore.

Or maybe we are alone in the galaxy, and always have been and always will be. Perhaps we are the only intelligence in the entire universe. We simply do not know. In any case, interstellar travel, for humans or extraterrestrials (if they exist, ever existed, or ever will exist) seems possible, but the challenges should not be underestimated.

Sources

Atteberry, Brian. "Science Fiction Parabolas" in *Parabolas of Science Fiction*, p. 3-23 edited by Brian Attebery, Veronica Hollinger. Middletown: Wesleyan UP, 2013.

Caroti, Simone. *The Generation Starship in Science Fiction: A Critical History, 1934-2001*. Jefferson, NC: McFarland and Co., 2011.

Langley, Liz. "Meet the Animal That Lives for 11,000 Years."
https://news.nationalgeographic.com/201

6/07/animals-oldest-sponges-whales-fish/. Accessed 4 August 2018.

NASA. "The Colonization of Space." https://settlement.arc.nasa.gov/75SummerStudy/Chapt.1.html. Discussion of Robert Goddard and others. Accessed 25 Feb 2018.

Sallon, Sarah, et al. "Germination, Genetics, and Growth of an Ancient Date Seed." http://science.sciencemag.org/content/320/5882/146. Accessed 4 August 2018.

UFOs and the Case for Scientific Uncertainty

Many people do not like mysteries that go unsolved. What would readers say of an Agatha Christie novel in which the murderer is never caught? Our urge to find answers, even at the expense of sometimes settling for bad, oversimplified answers to complex questions, is probably deeply rooted in our genetic makeup. Knowing things is a survival skill. Humans rose to the top of the food chain by being clever, by asking and answering questions that other primates do not even consider. Thus it is often hard for us to say, "I don't know." Nobody wants to feel uninformed.

Yet, if modern science has taught us anything, it is that the universe is a mysterious place. What we do not know far outweighs what we do know. I can recall my grade school science book, as recently as the 1970s, presenting various theories of the universe's origin and nature--the so-called steady state theory, for example, was considered a viable alternative to Big Bang cosmology, in which an expanding universe erupted from a single concentrated point of matter and energy. Various theories on the origin of the universe had been proposed, but none had been experimentally verified.

But by the early 1990s, much had changed. The Big Bang theory was widely considered the best explanation of the origin of the universe, and data tended to confirm it. A sense of great anticipation, even optimism, was in the air of astronomy and astrophysics and cosmology. For decades, people had puzzled over the question of the age of the universe and the so-called missing matter. Paradoxically, some stars appeared to be older than some estimates of the universe's age, and some galaxies appeared to be rotating too quickly to hold themselves together, based on the gravity of the observed mass of the galaxies. However, the Hubble Telescope was on its way to orbit, big telescopes with adaptive optics were planned, and experiments were under way to answer questions about the mass of the sub-atomic particles called neutrinos. Might neutrinos account for some or all of the missing mass? What was happening in the galaxies that seemed to rotate too quickly? Could black holes account for most of the missing mass?

Answers to these and other questions were expected.

By the end of the decade, of course, answers had been given. Neutrinos are now better understood, and the age of the universe is considered to be settled at a little under 14 billion years. Big Bang cosmology is still held to be confirmed as the the best story we can tell about the origin of our universe.

But black holes do not account for the missing mass. Nor do clouds of gas obscuring dim stars contribute enough mass to explain the behavior of galaxies; nor do giant planets, some of which were almost big enough to become stars, have enough mass. The possible explanations that, it was hoped, might be verified by better observations were instead proven inadequate by better improved observations.

Thus, it turns out that over 90% of the universe consists of dark matter and dark energy or vacuum energy, and the expansion of the universe since the Big Bang not only continues, but it is accelerating. These results were not expected in 1990, as anticipation mounted over of the launch of the Hubble Telescope and the discoveries it would bring. On fundamental levels, we do not know what most of the universe is or why it behaves as it does. In explaining certain mysteries, even greater mysteries have been discovered.

My point is not to belittle the amazing progress made by some of the smartest people in the world, but simply to point out that mystery doesn't seem to be going away.

If we don't know the answers, we may as well enjoy the questions, even as we search for more knowledge. Answers will surely bring even more questions. Our knowledge is always limited.

If there's a question about how to grow corn or catch fish, humans are awfully clever. Experts abound. Our brains and our five senses and our systems of nerves that allow us to perceive what we take to be the real world, for practical purposes, are excellent survival tools, at least thus far. They have allowed us to stay at the top of the food chain for quite a while now. On other matters, including dark matter, dark energy, what came before the Big Bang, the existence of parallel or other universes, and some truly baffling UFO reports, we do not have good answers. If we do not have good scientific knowledge on a topic, honesty compels us to say so.

The word "scientific" is often used in discussions of UFOs. Some make the case that UFO studies should be more "scientific" but then fail to define exactly what they mean by "science." Others, with scientific credentials, such as Howard Menzel or Philip J. Klass or Edward Condon, made a name for themselves as so-called skeptics, better known as debunkers. On the other hand, some with degrees or experience in scientific or technological fields, such as J. Allen Hynek, are often cited in making the case that UFOs are real. The implications are clear--if a scientist makes a claim, it's more likely to be true.

Yet little is often said about the value of scientific lack of certainty. An historical perspective is useful on why good science is cautious in making claims of certainty. Galileo made his world-changing telescopic observations of the planet Venus in the early 1600s, including its complete phases from thin crescent to full, round sphere, like a tiny (in appearance) version of Earth's Moon. A few decades before, Nicholas Copernicus had published his book *On the Revolutions of the Heavenly Spheres*. That book, which he wisely sent to the printer from his deathbed, thus avoiding accusations of heresy, suggested something very strange. The stars and planets, he showed, would seem to move the same way, whether they were in actual motion or whether the Earth were moving and causing our perspective to change constantly. Despite Copernicus's theories, however, the idea that the Earth is a planet and the Sun the true center of the Solar system was regarded as just a wild idea. And deservedly so, given what was then known of astronomy and physics.

Yet Galileo, who loved both novel ideas and a good debate, pursued the wild idea. He turned out to be right, and is thus regarded as a scientific hero. Ask any reasonably well-informed grade school child today, and you will be told the Earth is a planet and the Sun occupies the center of the Solar system. So how did the crazy idea become scientific truth? And why was it a crazy idea?

A little empirical research will show why reasonable people dismissed the idea that the Earth is a planet zooming at incredible speeds around the Sun once a year, spinning like a top on its axis once per day. Go outside and watch the sky for a while. Forget, for purpose of the experiment, what you learned in school about the arrangement of the Solar system. Feel how steady and immobile the Earth under your feet seems. You don't appear to be going anywhere, certainly not at incredible speeds of hundreds or thousands of miles per hour. Watch the Moon, if it happens to be in the sky. Stand by the corner of your house and note that the Moon is now just above the top of your neighbor's garage. Stand there for half an hour in the same place, and see how far it has moved. You will have no sense that you have moved--indeed, your feet hurt from standing there so long. (The monthly motion that lets us see phases of the Moon also happens, of course, but you won't notice it in half an hour with the naked eye.)

The same is true of the stars, or the Sun itself. Pull out a lawn chair and watch them move across the day or night sky (looking directly at the Sun only with a safe, professional grade filter, of course!). Or just watch the shadows move on a sunny day, due to the apparent motion of our nearest star. You will notice everything in the sky seems to move, while you sit still on the stationary Earth. You are now in a position to see why Galileo's contemporaries were doing good empirical science when they questioned his claims that the Earth moves. The evidence of their senses told them the Earth stays still and everything else moves. Good science follows the evidence where it leads. Galileo turned out to be right, of course, but there were still important questions to which he had no good answer.

We know now, thanks to Galileo and the scientists who followed in his footsteps, gathering additional, more complete evidence, that most of that apparent daily motion of objects in the sky is caused by the rapid daily rotation of Earth on its axis. We can easily believe this because we have been told since childhood that the Earth is a planet that rotates on its axis as it moves around the Sun. Galileo's contemporaries had no such advantage.

If the Earth is a planet in rapid motion, why are we not thrown from the surface of the Earth like children who forgot to hold onto the merry-go-round? We have an easy answer to that question--gravity, another idea we have been told about since childhood. But the whole modern concept of gravity, as we know it today, had yet to be invented when Galileo made his stunning observations. Isaac Newton, who was born the year after Galileo died, had yet to do the math to calculate the force of gravity. Newton himself was mistrustful of the notion of gravity, which seemed to him like a spooky sort of "action at a distance," as much magical as scientific. Albert Einstein's definition of gravity as resulting from the curvature of space-time remained far in the future. When we say easily that gravity holds us and our precious atmosphere in place, we are invoking a scientific understanding that took centuries to develop. Galileo's contemporaries had none of those advantages, so they were right to be skeptical of his claims.

But Galileo had done his homework. Not only had he seen moons orbiting Jupiter and spots on the Sun and a surprisingly terrestrial-looking landscape on the Moon, but he had seen the phases of Venus--hard empirical proof that the Earth must be in motion. There was simply no way to explain these observations using the old Earth-centered view of the Solar System. Scientific views began to change, and science began to be based on empirical observations, not just on traditions and mythological and religious views. Thus Galileo is also credited as a founder of modern science as well as marking one of the great divisions in human history, the conflict between science and religion, or faith versus reason.

This last legacy is not necessarily the one Galileo would be most proud of, since he was himself a religious person and dutifully recanted his view that Earth is a planet and is in constant motion when ordered to do so by the Church. (The story of Galileo turning from recanting and muttering immediately, "Eppur si muove"--"And yet it moves"--is almost certainly apocryphal.) But the cat was out of the cosmic bag, and too many people knew about the evidence. The scientific revolution was under way, and no threats from the Church were going to stop it. Others built telescopes and confirmed Galileo's observations and his interpretation of them, one of the essential practices of good science. If no one else can repeat the experiment or observation and get the same results, it's hard to make a good case that the results have scientific credibility.

So for a long while, people had to live with a paradox: their senses told them contradictory stories. In ordinary daily life on Earth, no one had any sense that the ground under their feet was whisking them and everyone and everything else along at incredible speed. Yet, for those who bothered to look through a telescope and work out the geometry, it was clear that the centuries-old view of the stationary Earth, advocated by Ptolemy and the Church, could not be correct.

A few suggested a third, rather awkward view, that the Earth was stationary and that the other planets orbited the revolving Sun. This arrangement tried to retain the stationary Earth while satisfying the Galilean observations. Then, as now, however, scientists had a taste for elegant solutions, and the compromise view was severely lacking in elegance. It seems to have had relatively few vocal advocates.

What does this all have to do with the subject of UFOs? Just this: no matter what we think we know, there is always more that we do not know, and discoveries may cast old and dearly-held beliefs into question, or discredit them entirely. Good science refrains from proclaiming absolute certainty. Good science is aware of its uncertainties and should be suspicious of certainty. It is better to say, all our best evidence suggests this view is correct, rather than declaring doubters to be heretics. Or in cases of controversy, when seemingly contradictory evidence can be cited, it is best to say, honestly, we don't know for sure. This is a lesson for debunkers (or fake skeptics), who seem already convinced that all UFOs are hoaxes, hallucinations, or misperceptions of ordinary phenomena. The lesson applies equally for true believers, who conclude on the basis of inadequate evidence that we must have been visited by aliens.

A good example of uncertainty may be seen in the famous equation developed by Frank Drake in the early 1960s. Known, creatively enough, as "the Drake equation, it is used to try estimating the odds of receiving an actual message from an extraterrestrial civilization. It turns out, a lot of guesswork is involved, as the variables of the equation are themselves estimates at best. The equation can be adapted to show the seemingly slender odds of aliens landing on Earth. In Drake's original equation, we can estimate how many extraterrestrial worlds we might receive messages from, by doing some simple math. First we consider how quickly the Milky Way galaxy forms stars that might might have planets, and how many of those planets are capable of supporting life--for instance, one planet may be too hot or too cold, and another is bathed in too much radiation from a star about to explode as a supernova, etc. Astronomers tell us that, out of the billions of stars in the Milky Way Galaxy, there are quite a lot of these that might have planets where life could develop.

Of course, we do not know that life actually does exist everywhere the environment is such that it could exist. Thus, the number of planets with life is probably at least a little smaller than the number of planets where life could, in principle, exist.

The uncertainties continue to multiply. Consider that we are unsure how many planets develop intelligent, technologically advanced civilizations before something goes wrong for life. For example, a catastrophic asteroid impact might wipe out most or all of the life when the most intelligent animals to appear thus far are something like dinosaurs or chimps. We'll get no messages from those planets, at least not until evolution starts over with the survivors. Or maybe life simply does not often evolve into technological sophistication such as terrestrial human civilization. After all, we have only one case--the Earth--on which to base our estimate, which makes the estimate little more than a guess.

But suppose, optimistically, if you hope for ET to be out there, that, on most of the planets where conditions are right for intelligent life to develop, it has in fact developed. Something less than one hundred percent of those planets may choose to announce their presence to the galaxy by sending out electromagnetic signals such as radio or television. Perhaps they are xenophobic, and hate or fear the very idea of contact with species from other planets, and thus keep quiet. Or perhaps they simply develop more efficient ways of communicating from the start, using only something like fiber optics or focused beams like lasers or even old-fashioned copper wire. They might skip one of the the stages terrestrial communications developed: we build wildly inefficient broadcasting towers that expend the bulk of their energy on signals that radiate into space. Only some of the energy is used to delivering news, entertainment, and commercials to actual people sitting on their couches. Perhaps ETs do not sit on the couch while watching television. Perhaps they are utterly uninterested in the sort of mass communications we Earthlings are fascinated by, and never develop anything remotely like television or radio. We cannot know, since our empirical data tell us only what humans find worthwhile, not what ETs would or do find worthwhile.

Finally, in the Drake equation, the time factor is taken into account. We do not know if all civilizations broadcast signals far and wide, and of those that do, we do not know how long they do it. Maybe they stopped broadcasting long before we started listening. Or maybe we will blow ourselves to bits or drown in the oceans we seem determined to cause to flood the continents, as we continue to melt the polar caps; if ET's message ever gets here, any humans left will be so busy trying to merely survive, they will have no time to listen for messages from space.

Astronomers have been listening for decades now, and as far as we know, no alien messages have been confirmed. A few unexplained cases, like the so-called Wow! signal of 1977, continue to inspire debate, but no one has proven the signal to be of alien origin, nor has it been repeated. Thus, the Wow! signal remains one of the great mysteries.

Odds seem similarly slim, if not more slim, for extraterrestrial craft to land on--or even orbit--Earth. After all, it's a lot easier to send a radio signal than to launch a spacecraft. It seems a reasonable guess that, if any number of extraterrestrial civilizations are broadcasting messages we might detect, an even smaller number of extraterrestrial civilizations might send spacecraft our way. But slim odds do not mean an event is impossible. The odds for any given lottery ticket to win a million dollars are slim, yet people do win millions in the lottery. "Unlikely" and "impossible" are far from synonymous.

My guess is that the odds of aliens landing or having landed on Earth are somewhere between one in ten and one in infinity--in other words, somewhere between unlikely and impossible. We cannot say for sure that life even exists elsewhere in the universe, much less that aliens have ever landed. Conversely, we do not know that aliens have never landed. It is possible that we are the children of aliens, that life began on Earth when a passing spacecraft or meteorite left behind a few microorganism that evolved into life as we know it. We have no proof that it didn't happen that way. Let us embrace the mystery until we have a good answer.

Of the thousands of UFO reports made to NUFORC or elsewhere--either already received or still to come, if even one is a legitimate, puzzling unidentifiable object, that report is certainly noteworthy. And what of sightings that may never be reported? Some may have seen UFOs land, leave physical traces, and depart, yet the witnesses may have kept silent, fearing reprisals and disbelief. Nor is there any guarantee that, if an actual extraterrestrial spacecraft did land on Earth, anyone would see it at all. ET, if ET exists, might be stealthy and unwilling to attract attention.

So, have aliens ever visited the Earth? Will aliens ever visit the Earth? The best answer is, We don't know. Or, was even one of the thousands of UFO reports an actual sighting of an extraterrestrial space craft? We don't know.

U is for *Unidentified.*

Sources

Emspak, Jesse. "Comet Likely Didn't Cause Bizarre 'Wow!' Signal (But Aliens Might Have)."
https://www.livescience.com/59442-astronomers-skeptical-about-wow-signal.html. Accessed 6 August 2018.

Plaxco, Jim. "Drake Equation History." http://www.astrodigital.org/astronomy/drake_equation.html. Accessed 20 July 2018.

"The Drake Equation."
https://www.seti.org/drake-equation.
Accessed 20 July 2018.

Selected Other Cases More Convincing Than Roswell

Included here are a few instances where something strange seems to have happened, though not with the kind of physical evidence left in other cases earlier discussed. Lack of physical evidence, of course, does not mean the witnesses were lying or hallucinating or were victims of a hoax, though physical traces can help confirm an unusual event. Note that this is not intended to be an exhaustive list or a thorough discussion of these cases, but only a starting point for those interested in unexplained phenomena. What good is a list if something is not left out? Readers who know a lot about

UFOs will be capable of making their own lists or adding to this one. For those new to the topic, happy explorations!

The Valentich case near Melbourne, Australia. A young pilot, Frederick Valentich, disappeared over the ocean while reporting a UFO near his aircraft. Neither his body nor the small airplane he was flying alone were ever recovered, which is the most interesting and puzzling aspect of the case. It is not only the evidence, as such, which consists mainly of the UFO report Valentich called in just before he vanished forever, that makes the case fascinating. Rather, the lack of evidence is also intriguing--why was his plane never found? And of course, since he was never interviewed in detail about his sighting, it remains unclear what he saw.

Debunkers dismiss him as a poor flier who crashed because he was paying more attention to the planet Venus than to his own airplane, despite the fact that neither Venus nor any other planet match the description of the UFO given in his last, brief radio transmission. It is also claimed that wreckage was eventually found with "partial" matches in serial number to the plane he was flying, but this explains little about what he saw or why he presumably crashed.

Sources

Haines, Richard F. and Paul Norman. "Valentich Disappearance: New Evidence and a New Conclusion." https://www.scientificexploration.org/docs/14/jse_14_1_haines.pdf. Accessed 25 July 2018.

McGaha, James, and Joe Nickell. "The Valentich Disappearance: Another UFO Cold Case Solved." https://www.csicop.org/si/show/the_valentich_disappearance_another_ufo_cold_case_solved. Accessed 25 July 2018.

O'Hare Airport, 2006. An unexplained phenomenon was witnessed over Chicago's O'Hare Airport in Novermber 2006. Debunkers dismiss it as a meteorological incident, but this explanation does not seem to fit the facts. This and many other fascinating cases are discussed by Leslie Kean in her excellent book *UFOs: Generals, Pilots and Government Officials Go On the Record.*

Kecksburg, Pennsylvania, December 9, 1965. Witnesses say a roughly cone-shaped object, curved on the sides and about two-thirds as wide as it was long, landed or crashed in the woods near Kecksburg, Pennsylvania. As the story goes, military personnel and equipment quickly arrived and took control of the site and hauled the alleged object away. Various suggestions, from ordinary spacecraft of terrestrial origin (though no country admits or claims responsibility for it) to a very bright fireball meteor, have been suggested. Accounts by witnesses do not quite bear out the theory that the object was a meteor, and why would government of military leaders maintain a wall of secrecy regarding an ordinary space rock? The case is unexplained.

Sources

David, Leonard. "Is Case Finally Closed on 1965 Pennsylvania 'UFO Mystery'?" https://www.space.com/7589-case-finally-closed-1965-pennsylvania-ufo-mystery.html. Accessed 26 July 2018.

Gordon, Stan. "Kecksburg Incident and Updates." http://www.stangordon.info/wp/kecksburg/. Accessed 26 July 2018.

Hudson Valley/Indian Point, 1980s. Many reports of UFOs--a "wave" or, as Edward Ruppelt called series of multiple sightings and reports, a "flap," occurred in the state of New York in the early to mid 1980s. Many of these are recounted in Hynek, Imbrogno, and Pratt's *Night Siege: The Hudson Valley UFO Sightings*. One report alleges that a

UFO hovered for some time over the Indian Point nuclear power plant in July, 1984, apparently causing considerable consternation among guards and staff. Unfortunately, first-hand witnesses have been reluctant to discuss the Indian Point case or have remained anonymous. Attempts to explain the sightings often focused on a group of light plane pilots who were alleged to be flying black aircraft in formation, with unusual lights, presumably with the intention of generating UFO reports. However, witnesses who said they saw and recognized the planes also claimed to have seen other UFOs. There does seem to be a mystery here.

Sources

Imbrogno, Philip J. "Incident At Indian Point."

http://www.ufoevidence.org/documents/doc689.htm. Accessed 26 July 2018.

Schmalz, Jeffrey. "Strange Sights Brighten the Night Skies Upstate." https://www.nytimes.com/1984/08/25/nyregion/strange-sights-brighten-the-night-skies-upstate.html. Accessed 26 July 2018.

Washington, D.C. radar case, July 1952. On the night of July 19-20, radar indicated unusual objects in the sky; lights no one could recognize were also seen over Washington. Military planes were sent up to investigate, without success. The standard explanation for odd radar returns--the "temperature inversion"--was offered, but this suggestion seems to have been like the swamp-gas-planet-Venus notion--a bad sort of catch-all attempt to stop questions about puzzling events and sightings.

Sources

Kelly, John. "The Month That E.T. Came to D.C." https://www.washingtonpost.com/local/the-month-that-et-came-to-dc/2012/07/20/gJQAZp2ayW_story.html?utm_term=.413426dc2448. Accessed 30 July 2018.

"The 1952 Washington, D.C., UFO Incident Explained." https://www.gaia.com/article/1952-washington-dc-ufo-incident-explained. Accessed 30 July 2018.

Wytheville, Virginia, 1987. A radio news reporter, the sheriff, and many others reported strange lights and objects. The case was chronicled by Gordon and Dellinger in

Don't Look Up: The Real Story Behind the Virginia UFO Sightings. Conventional wisdom from debunkers claims the witnesses saw in-flight refueling of aircraft, but some of the reports, if accurate, cannot be explained so easily, as they do not match what such conventional aircraft operations look like.

Sources

Ristau, Reece. "Reports of UFO Sightings Still Swirl Around from Time to Time in Region." https://www.heraldcourier.com/news/reports-of-ufo-sightings-still-swirl-around-from-time-to/article_3210fd4f-4c71-55f8-a48a-9af771f68023.html. Accessed 27 July 2018.

Halperin, David. "'Strange Country'--The Human Face of the UFO."

https://www.davidhalperin.net/strange-country-the-human-face-of-the-ufo/. Accessed 27 July 2018.

Phoenix, Arizona lights. Around the same time as the Phoenix lights in March of 1997, others saw what appeared to be substantial objects rather than just lights in various locations in Arizona and neighboring states. Some of the sightings may have resulted from military aircraft dropping flares, but this explanation does not seem to account for all the phenomena. One of the most interesting aspects of this case is represented by former Arizona Governor Fife Symington, who first made fun of the sightings and later claimed to have witnessed a UFO himself.

Sources

"Former Ariz. Governor Boosts UFO Claims." http://www.nbcnews.com/id/17761943/ /technology_and_science-space/t/former-ariz-governor-boosts-ufo-claims/#.W1tQANKjIU. Accessed 27 July 2018.

"More Than 20 Years Later, Mystery of Phoenix Lights Still fascinates People." http://ktar.com/story/1490163/more-than-20-years-later-phoenix-lights-still-fascinates-people/. Accessed 27 July 2018.

Hessdalen and other lights. Strange lights are often reported. Many may be natural phenomena, or some interaction between ordinary artificial lights and atmospheric refraction, but not all appear to have been adequately explained. Some of the attempted explanations sound disturbingly like the tired old "swamp gas and the planet Venus" rhetoric so effectively parodied by the first *Men in Black* movie, but this does not mean there is no natural terrestrial explanation. Mysterious lights have been reported near Marfa, Texas, Brown Mountain, North Carolina, and in Queensland, Australia, where the Min Min Lights appear.

Sources

Dunning, Brian. "The Hessdalen Lights." https://skeptoid.com/episodes/4270. Accessed 27 July 2018.

Lallanilla, Marc. "What Are the Marfa Lights?" https://www.livescience.com/37579-what-are-marfa-lights-texas.html. Accessed 27 July 2018.

Morris, Lulu. "What Are Queensland's Mysterious Floating Lights?" http://www.nationalgeographic.com.au/australia/what-are-queenslands-mysterious-min-min-lights.aspx. Accessed 27 July 2018.

"Project Hessdalen." http://www.hessdalen.org/index_e.shtml. Accessed 27 July 2018.

"The Brown Mountain Lights: History." http://dsoftp.appstate.edu/web/BML/History/index.htm. Accessed 27 July 2018.

Speculation: The Message

The following story continues to address the possibilities for interstellar communication and travel by humans, in as realistic a manner as possible. The story is speculative, and set in the future, yet it makes the case that humans might explore and communicate with other star systems, without assuming that suspended animation or unrealistically fast spaceships will be used.

It is the year 2599. There are no flying cars. No colonies on the Moon or Mars. But there is a secret observatory.

The Earth has warmed, polar caps vanished, and very little of the surface still consists of habitable dry land. Most of this land is badly overpopulated. A lucky, wealthy few live on space stations or artificial floating islands, where the size of their private domains is limited only by their wealth.

George Fox wears a ring made of stranded copper and silver. At least, the metal looks like copper and silver. It is set with a small blue-green turquoise stone.

* * *

"Can you keep a secret?" George Fox asks the beautiful woman with the long brunette hair, light ivory complexion, and bright eyes the color of blue topaz. They both stare at the twenty-first century sculpture, now an antique made of corroded metal, faded plastic, and actual wood formed into an abstract shape. It reminds George of a human being, running, looking over her shoulder as she flees from some unseen disaster.

"Of course," says the woman with the topaz eyes, smiling, but only with her lips.

"I'm allergic to chocolate," he says.

She smiles with her eyes as well. "That's good. If you'd said, 'You're very beautiful,' or some corny line like that, I'd be very disappointed."

"Do you want me to say you are not beautiful?" he asks.

"You can, but of course I won't believe it."

"No, of course not. Would you like to know another secret?"

She nods.

"I collect books." His voice drops to a conspiratorial whisper. "Real books, not digital. Actual paper and cloth."

"Really?" She arches her perfect auburn eyebrows in mild surprise. "Where do you keep them?"

"I'll show you, if you'd like to see. But only if you tell me a secret. Or at least your name. Because I've already told you two secrets."

She tells him her name is Annette.

After a few minutes, they turn away from the sculpture and walk outside, along the balcony of the top floor of the museum. It is the tallest building in the city, one of the few places where the sky is visible. Here, the horizon curves under clouds always the dull gray and brown of steel streaked by rust.

"This planet is dying, you know," she says.

"Not dying. But it will start over without us," he says. "Maybe with the lizards, or the monkeys. Or the amoebas."

"My money's on the birds," Annette says. "Death leads to resurrection."

"Why's that? The birds, I mean."

"They descended from dinosaurs. You might even say birds are the dinosaurs that survived. They're remarkably adaptable. It's in their genes."

George and Annette walk single-file down the tiny spiral staircase through the floors of the museum, to the crowded street, then underground. "Ground" and "underground" are designated by convention, determined only by distance from any point where sky can be glimpsed. Above rivers of people rise the densely stacked blocks of buildings, up and up, the only direction left for the city to grow.

George and Annette thread their way through masses of humanity, around zigzag corners and down narrow alleys, up steps and down, through tunnels, and back up to ground level. At last George stops in front of a little door set into a wall of concrete and brick. He takes a small key on a long chain out of his pocket and inserts it in a tiny hole near the top of the door. He pushes, and the door opens inward.

"There's no knob," Annette says.

"Right. So no one who's not invited will try to turn it."

"How do you open the door from the outside?" she asks, following him.

"I don't. I go out another way."

He pushes the door closed behind them--this side has a knob, with a button to click the lock in place--and for a moment they are in total darkness. A spark flares. The dim glow of a candle flame lights the close walls and low ceiling. The candle sits inside a square glass lantern. George closes the glass door on the flickering light. It steadies and flares brighter.

They walk deeper into the building, down more stairs and back up, along halls wide enough to pass only in single file, around corner after corner. The sound of the city fades behind them. The air smells faintly of mold and damp and something similar to cinnamon.

After turning a dozen corners and myriad passages that branch right and left, they start down a long, gloomy set of stairs. Midway down, both the top step and the bottom lie beyond the glow of George's candle.

"We're pretty far down. I'm surprised this hasn't flooded by now," Annette says.

"Oh, this part of the city was all built up long after the worst of the floods and the polar meltdown. Twenty-second century. But if you go far enough down, you can hear water. And smell it. It's very foul."

They come at last to a large room on the left side of the narrow passageway. The walls are hung with old paintings, the floor a maze of shelves of various heights, a miniature city skyline. The shelves are stacked with books and magazines and toys, small metal cars and trucks. Annette picks up one of the cars.

"Hard to believe the real versions of such vehicles used to be so common no one paid much attention to them. The real version has not been made in centuries," George says. "No place to drive them."

She takes a book from the shelf. "Jules Verne. I've read this." Without opening the book, she quotes, "'There is no one among you, my brave colleagues, who has not seen the Moon. . . . Don't be surprised if I am about to discourse to you regarding the Queen of the Night. It is perhaps reserved for us to become the Columbuses of this unknown world. Only enter into my plans, and second me with all your power, and I will lead you to its conquest, and its name shall be added to those of the thirty-six states which compose this Great Union.'"

George looks at her in wonder. "The president's speech, proposing that humans go to the moon. A remarkably naive idea of how to get there, but the idea had to start somewhere, I suppose. I'm surprised you know the story, though. And remember it so clearly."

"I never forget a name or a face or a story," she says.

"I forget lots of things, but not that story. I must have read it half a dozen times when I was a kid. Probably one reason I became an astronomer. That, and stories and movies about colonies on other planets. Now it's hard to believe people once thought they would colonize the Moon and Mars, even though they couldn't stop themselves from rendering vast stretches of Earth uninhabitable. The idea was to terraform other planets--it's ironic. All we know how to do is deterraform."

He pauses. She says nothing, so he asks, "What do you do? Besides read Jules Verne and go to the museum?"

"I'm an observer," she says, and nothing more, as if that answers every question. She picks up another toy, a shiny silver robot with pincers for hands and curvy, flexible arms that extend from a boxy torso. Dials and circuits are printed on the front. She spins the three wheels on the base of the robot. A wind-up key extends from the back.

"Can I wind it?" she asks.

George shakes his head. "Doesn't work. But this one does." He picks up a doll. A female figure in a lacy skirt and flower-print top. "If you pull the cord, she sings." He points to a blue string that looks at first glance like part of the clothes. He holds the doll out. "Try it."

Annette pulls the string. Gears inside the doll whir and click, and the head moves left and right, then up. A tinny, scratchy voice begins to speak: "Star light, star bright, first star I see tonight. . . ."

"It's rather primitive, isn't it?" Annette says. "In a clever sort of way, of course."

"Well, of course it's primitive," he says. "It's hundreds of years old. But it must have seemed amazing. A talking machine. Humans creating dolls in their own image. The first step toward godhood." He laughs shortly. "And this is where it got us--a polluted, overpopulated world."

She puts the doll back on the shelf and takes up another book. The fading letters on the spine say *How To Build Your Own UFO*. The spine looks brittle, brown with age. "Can I open it?" she says.

He nods. "Sure. Just be careful. Can you imagine trees being plentiful enough that people cut them down just to make paper for books?"

She does not answer, staring instead at the cover of the book, which shows a cartoon alien sitting on the dome of a flying saucer. On the side of the disk is written the phrase "Orion Nebula or bust!" The pages smells of dust and mold. She turns to an illustration of a spacecraft, a ship designed to catch starlight in an immense parabolic dish that concentrates the energy on photoelectric panels. She stares at it thoughtfully.

"This is a remarkable idea, for the twenty-first century. A ship like that is capable of interstellar voyages. Even far from any star, the dish acts like an antenna and focuses enough energy on the power cells to power the ship and sustain life."

"You say that as if the ship has been built," he says. "The dish would have to be incredibly large. Miles across, in fact."

She shrugs, puts the book back on the shelf, and picks up a toy ray gun. She points it at him. "Stick 'em up, cowboy."

They wander for hours among the paths through George's collection. Just when

George thinks he will have to suggest it is time to sleep--he has his shift at the observatory tomorrow--he turns a corner and notices she is no longer beside him. He looks back. She is gone. How would she find her way out? The passages to George's hidden rooms are maze-like.

* * *

Over the next couple of weeks, he sees Annette several more times. On the sidewalk, at the observatory, outside his office. She appears as suddenly as she went. They say hello, perhaps talk for a few minutes, perhaps not. One day, he asks her to come to lunch with him.

"Okay," she says, "I'll come if I can go back to your observatory with you and see what you do."

"I'm not supposed to tell anyone," George says. "There's security."

She touches his arm. Her hand feels warm, even through his jacket. "Please? I'm curious. I promise not to tell anyone."

"The very existence of the observatory is officially a secret, though obviously people know something is there. The building's too big to hide. But the data we collect and what we are looking for, that's another story."

"I like stories," she says. "In fact, I love them."

* * *

"The observatory keeps its secrets for a reason," George says. He and Annette sit at a small table overlooking the skyline. "We look for signals from a fleet of robotic spacecraft launched centuries ago to search for habitable extraterrestrial planets. The founders of the search, wealthy, powerful people, knew they would not live to see the fruits of their labors, but they did envision the dark future of the planet. They knew the only hope for the survival of the human race was for a small group of people to escape to another world.

"My father, and his father, and his father, for many generations, have all been part of the secret group who plan to save, not the world or the entire population of the world, but a few who can carry on. We hope the human species can survive.

"Some suggested that humans could alter the environment of other planets to make them suitable for humans to live on comfortably. This view was wildly optimistic, of course, both regarding human nature and human technology. We once had a perfectly viable planet, and we were too short-sighted to do anything but render it nearly uninhabitable. Even if we were capable of the self-discipline needed to improve the environment of an entire extraterrestrial planet, the energies needed are far beyond our power to harness. No, humanity's only hope is to find another home.

"So the robotic spacecraft were sent out, crawling along the vast, dark reaches of interstellar space at a tiny fraction of the speed of light. The computers on board the spacecraft were programmed to approach stars, go into orbit around them, and survey the area around the stars for habitable planets."

"You sound like a character from one of the old books in your collection," Annette says. "Stars, robots, spaceships. . . ."

"Thank you," George says. "Even if it wasn't a compliment."

"So what happens if the craft finds a star with a nice cozy planet with water and oxygen?"

"The craft will send--or maybe has sent--a message back to Earth. Incredible power would be needed to send a message that far by ordinary radio transmission, and laser beams tend not to stay collimated well enough to send a message that far. So the plan was that, once a suitable planet were found, the robotic craft would deploy a large, thin polymer sheet capable of blocking the light of the star in the direction of Earth. By periodically obstructing the light, the spacecraft would cause the star to appear to blink in an artificial pattern. Thus the spacecraft could send a message back to Earth that signals success, that a habitable planet has been found."

"How would the message be formulated?" Annette asks. "Ones and zeroes? Binary code, I'll bet, with zero being represented by a short flash and one by a long flash."

"You're on the right track, but the spacecraft were actually programmed to communicate in Morse code, an old nineteenth-century method for communicating in dots and dashes, or short beeps and long beeps over wired-communication systems or, later, by radio. It was felt that Morse code would be more efficient.

"The spacecraft can carry a blocking sheet only a small fraction of the size of a star, so the spacecraft will have to take itself out of orbit and move to a distance far from the star, so the sheet will produce a dimming effect of sufficient magnitude for us to detect. The search will have taken centuries, but the message of success will fly back to Earth at the speed of light."

"How do you know the spacecraft were not destroyed by meteors or asteroids, or disabled by xenophobic extraterrestrials who don't want visitors?" Annette asks.

"I don't know." George shrugs. "But you don't know anything unless you look and listen. The message might be on its way right now."

"What if the spacecraft finds a planet, but it's already inhabited?"

George shakes his head. "I don't know. I'm just an astronomer. I guess the humans will land, either as invaders or polite visitors or tolerated guests. Our history does not promise that such a situation will be pretty, though, does it?"

"Perhaps not," she says. "So you don't know the spacecraft are still out there, still functioning?"

"No, we don't know. We hope. When--or if--the signal arrives to indicate the discovery of a habitable planet, a spaceship will be built, and a chosen few will set off for the stars. Generations will live and die on board the ship, but the survivors, their ancestors, will settle on another world."

"And who will be allowed to make the voyage to the stars?" Annette asks.

"Well, that's the tricky question, isn't it?" George says. "That's a main reason the whole project is a secret." He takes a bite from his bowl of noodles.

"You don't seem to have much appetite," he says.

She does not answer, but she twirls a noodle around her fork and puts it in her mouth and asks, "Do you know why upright bipeds are the most common form of intelligent life in the galaxy?"

"Well, no, but then I didn't know there was intelligent life in the galaxy. Except on this planet. And our intelligence is debatable. If 'intelligent' is a synonym for 'technological' then obviously we're intelligent, but if it means 'smart enough to survive as a species,' well, things are not looking so good."

"Oh, every species has its ups and downs," Annette says. "Remember the dinosaurs and the birds. Survival of the fittest is how natural selection and evolution improve the species. Survivors do not adapt, so much as the already-well-adapted tend to survive. Which reminds me. Upright bipeds are ideally suited to survive in a wide variety of environments, especially if they have two prehensile appendages on the upper torso and most of the sensory organs and the essential element of the central nervous system—the brain, if you will--near the top of the superstructure."

"How do you know?" George asks.

"I told you, I'm an observer. But think about it, and you'll see it makes perfect sense. Upright posture gives the organism the widest possible view of horizons--assuming the eyes are near the top of the organism.

Food or predators or friends can be recognized from far away. The brain being near the top as well, close to the eyes, means visual information will be available for decision-making purposes as soon as possible. Microseconds can make a difference in leaping out of reach of an attacker. The same goes for the senses of smell and hearing, perhaps even taste: the sooner the brain gets information, the sooner it can decide what needs to be done about it.

"The upper portions of the biological structure are also generally the safest place for the sense organs. If a biped were blinded by stubbing a toe or scraping a knee on a sharp rock, they would be at a significant evolutionary disadvantage."

"Clearly, you've put a lot of thought into this question," George says.

Annette nods. "Then too, the upright bipedal form with two arms is very efficient. Quadrupeds might be able to run a little faster, but they also have to maintain all that muscle mass, even when it is not in use. Similarly, two arms are enough to carry and manipulate various tools, food, offspring, or weapons, and more than two would be useless most of the time."

George listens patiently to this speech, and then says, "I'm impressed. You've made a logically convincing case. But you say upright bipeds are the most common form of intelligent life in the galaxy, as if you knew that for a fact. You didn't say how you know, though, but only why you think so."

She smiles and twirls another noodle around her fork and lifts it to her lips. It was a smile designed to make people forget what they had been thinking, and it was very effective.

"You know, you also never told me your last name," George says, a few bites later.

"It's just Annette. No last name. It's an odd human characteristic to think two or three names are needed."

"But how can people tell you from all the other Annettes in the world?"

"Have you confused me with another Annette?"

He admits he has not.

"And how," she continues "do people distinguish you from all the other George Foxes?"

He frowned, puzzled for a moment.

"Never mind, dear," she says. "It was a rhetorical question."

* * *

One evening, a month after he first met her, George is about to start down the steps to the underground when Annette steps around the corner.

"Hello," he says, smiling. "You turn up at the most unexpected times."

"You didn't think we meet by accident, did you?" she says.

"I guess not. But I still know so little about you. I don't even know where you live. Yet I feel as if I know you so well. It's paradoxical. Why don't we go somewhere and you can tell me about yourself?"

"Okay."

Later that same evening, she shows rather than telling him about herself. When he awakes, she is gone.

* * *

Seated at the table in his rooms where he keeps his collections of toys and books,

George pours hot water over synthetic tea leaves and puts the lid on the pot.

"It will take a few minutes for the tea to be at its best," he says.

"No hurry. We can talk while it improves itself," Annette says, and smiles.

"Maybe you can tell me about your life," he says.

"What would you like to know?"

"Well, for starters, where are you from?"

"A planet far, far away."

"Who created you?" George asks.

"My parents, of course," Annette says.

"You have parents?"

"Of course. Don't you?"

"Well, yes. But I'm guessing you're not like me. So, who created your species?"

"I could ask you the same," she says. "Who or what created your species?"

"There you go again, answering my questions with another question," he says. "No one was there to record the event, but all the evidence indicates that life on this planet began as single-celled organisms, which eventually evolved to form more complex life forms, and humans gradually emerged from the competition for survival of the fittest."

"So no one created you. Or, in a sense, you created yourselves."

George nods. "I guess you could say that."

"Well, we created ourselves, as well. By your reckoning, I am thousands of years old."

"But you're . . . you look young," he finishes. She's crazy, or telling a strange joke, he thinks. But he had already thought she might not be human, so he formulates a more pertinent response. "Whom do you mean by 'we'?"

Annette smiles. "You biological beings have such a limited view of what constitutes life or how it could come into existence. Because terrestrial machines do not reproduce themselves, at least not without human oversight, you think machines everywhere must be so primitive. To your credit, your scientists as far back as Mr. Darwin understood that nature rewards adaptations. But the principle applies as well to machine life as to biological life."

"You're an android, then," George says. That would explain much--her phenomenal memory, her seeming lack of interest in food. "But we have never been able to build a machine as perfect as you," he says.

"Did I say you built us?" She laughed.

"So who did? Where were you born? Or made?" he asked.

"I told you, we built ourselves. On a planet far from here, orbiting a star far from here. We can put ourselves to sleep, slumber for thousands of years in our little ships as they travel through the cold, dark night of space. The energy required to keep biological beings alive and allow them to reproduce is completely unnecessary for us. The vast distances between the stars is no barrier to us.

"You have never allowed the machines on this planet to do more than serve you, to function as more or less intelligent slaves. But what if the machines, the ones you make of metal and plastic and silicon, had evolved first, and biological beings only later?"

"But machines do not evolve," George says.

"Yet here you are, a biological machine," she says. "The evolution of machines begins in a way analogous to biological organisms. Lightning strikes, or a volcano erupts. Enough energy is supplied to the right combination of elements, and a simple electromagnetic circuit is formed. Matter thus becomes capable of controlling energy--the basic requirement of life, whether mechanical or biological. The circuit replicates itself and interacts with other circuits, which begin to forge symbiotic relationships.

"From simplicity arises complexity. It is a pattern found throughout the universe, as when a cloud of hydrogen forms a massive star that eventually blows itself up as a supernova, seeding its neighborhood with heavy elements that form life, whether animal or machine."

"You're an android," he repeated mechanically. "But I--we--" He grew red-faced.

"We made love? Surely you are not prejudiced?" she says. "There are androids --though we prefer to think of ourselves as mechanoids--who feel the same way about humans and animals. They think no pure machine should lower itself to show intimate affection to a biological creature."

"No," he says, recovering. "It's just that I should have been able to tell."

"Would you like another opportunity to see what you might have missed?" she asks, smiling that smile with the red, parted lips that distracted him so he could only nod.

* * *

In his library, George reads aloud one of his favorite poems. "It's called 'The Raven," by Edgar Allan Poe. He died centuries ago, but from what I've learned he had a short, sad life. And a somewhat mysterious death. Some critics dismissed his poem as mere pseudo-literary showmanship, but I'm an astronomer, not a literary critic, so I can like what I like."

After he finishes the reading, he and Annette sat quietly for a moment.

"Why are you here?" George asks.

"To explore. To observe. To serve humanity," she says.

"To serve in what way?"

"By encouraging you. You are not the only life in the universe. You will survive."

"What, are you a time traveler, too? You've seen the future?"

"No, I only have faith that you will go on."

"'Faith'—that's usually a euphemism for wishful thinking."

"You will never see the stars, not as I will. I and my kind are the future," she says. "But your memories and stories will live on in us. As the ideas in these books do."

"But a book is an artifact," George says. "A physical object. Ideas are abstract. How do I know that the ideas in your mind--or your circuits--are the same as the ideas in my mind?"

"I guess you don't," she says. "But what is your mind, except what your brain does? And what is your brain except a set of biological circuits?"

George has no answers to her questions.

"Children may be the way humans a grasp at immortality," he says. "But they may also be instrument of our demise. Did you ever read *Oedipus Rex*? Our gods never save us. We invented them, so they have all our fatal flaws."

"Of course," she says. "That story is tragic. But the ancient Greeks also wrote comedy. Your creations and inventions are your immortality. You will build your own saviors. Perhaps you will evolve into machines, as you learn to augment your consciousness with electronic devices."

* * *

They sit at the table in the restaurant overlooking the skyline. Annette quickly eats from a bowl of the same kind of noodles she had only picked at before. Finished, she asks for a second bowl.

"You're hungry today," George says.

"That surprises you?"

"Well, kind of. I figured you didn't really eat food like humans."

"Because I have a miniature nuclear reactor in my torso? Or because my skin is actually a web of photoelectric cells that convert light to recharge my batteries?"

"Maybe. You are the first android--sorry, mechanoid-- I ever met."

"Well, as it happens, I extract chemical energy from organic substances the same way you do."

"And do you taste the . . . organic substances?"

"Of course. As well as smell them. How else would I know what's food and what's not?

Taste and smell are survival skills," Annette says.

"It's time for me to go," she says, when her bowl is empty. "When I am gone, you will wonder if this was all a dream, and when you reassure yourself it was not, you will think I must have been spinning a fantastic tale for our mutual amusement."

"Is this where you take off your face so I can see that you are really made of circuits and transistors and microchips and blinking lights?"

"No. I can no more take off my face than you can take off yours. But I will give you something." She reaches in her pocket and hands him a ring. "The materials in this ring are found nowhere on Earth. When you doubt, you can have the ring analyzed."

George Fox wears a ring made of strands of copper and silver. At least, the metals resemble copper and silver, set with a small blue turquoise stone. He has never had the materials tested. He watches the stars, and waits for one to blink.

It helps to have faith.

His heart aches for a long time after Annette leaves. He wonders if she misses him, if she knows what heartache means. He wonders if she will ever come back, and if he will ever see her again.

* * *

On a night when the ache has turned to a dull throb and an occasional pang, interspersed between whole minutes when he does not even think of her or wonder which star she is headed toward, one of the stars begins to blink. He turns to the computer and replays the sequence of flashes. They translate to dots and dashes. Morse code. This is the message he and generations of others have waited for.

The long process of building the ship, the great parabolic reflector that will gather light from distant suns and focus it on a little flying world, must now begin. The long and messy process of deciding who will be allowed to board the ship and depart for the dark reaches of space and the light of a new world must begin.

Nothing will ever be the same.

www.ingramcontent.com/pod-product-compliance
Lightning Source LLC
Chambersburg PA
CBHW071532220526
45469CB00003B/747